VOLUME FORTY THREE

ADVANCES IN CHEMICAL ENGINEERING

Control and Optimisation of Process Systems

ADVANCES IN CHEMICAL ENGINEERING

VOLUME FORTY THREE

ADVANCES IN CHEMICAL ENGINEERING

Control and Optimisation of Process Systems

Edited by

S. PUSHPAVANAM
Chemical Engineering Department,
I.I.T Madras,
Chennai 600036

ELSEVIER

AMSTERDAM • BOSTON • HEIDELBERG • LONDON
NEW YORK • OXFORD • PARIS • SAN DIEGO
SAN FRANCISCO • SINGAPORE • SYDNEY • TOKYO
Academic Press is an imprint of Elsevier

Academic Press is an imprint of Elsevier
525 B Street, Suite 1900, San Diego, CA 92101–4495, USA
225 Wyman Street, Waltham, MA 02451, USA
32, Jamestown Road, London NW1 7BY, UK
The Boulevard, Langford Lane, Kidlington, Oxford, OX5 1GB, UK
Radarweg 29, PO Box 211, 1000 AE Amsterdam, The Netherlands

First edition 2013

ISBN: 978-0-12-396524-0
ISSN: 0065-2377

For information on all Academic Press publications
visit our website at www.store.elsevier.com

CONTENTS

CONTRIBUTORS

Dominique Bonvin
Laboratoire d'Automatique, Ecole Polytechnique Fédérale de Lausanne, EPFL, Lausanne, Switzerland

Grégory Francois
Laboratoire d'Automatique, Ecole Polytechnique Fédérale de Lausanne, EPFL, Lausanne, Switzerland

Sanjeev Garg
Department of Chemical Engineering, Indian Institute of Technology, Kanpur, Uttar Pradesh, India

Santosh K. Gupta
Department of Chemical Engineering, Indian Institute of Technology, Kanpur, Uttar Pradesh, and University of Petroleum and Energy Studies (UPES), Dehradun, Uttarakhand, India

Wolfgang Marquardt
Aachener Verfahrenstechnik - Process Systems Engineering, RWTH Aachen University, Aachen, Germany

Adel Mhamdi
Aachener Verfahrenstechnik - Process Systems Engineering, RWTH Aachen University, Aachen, Germany

Siddhartha Mukhopadhyay
Bhabha Atomic Research Centre, Control Instrumentation Division, Mumbai, India

Arun K. Tangirala
Department of Chemical Engineering, IIT Madras, Chennai, Tamil Nadu, India

Akhilanand P. Tiwari
Bhabha Atomic Research Centre, Reactor Control Division, Mumbai, India

PREFACE

This issue of *Advances in Chemical Engineering* has four articles on the theme "Control and Optimization of Process Systems." Systems engineering is a very powerful approach to analyze behavior of processes in chemical plants. It helps understand the intricacies of the interactions between the different variables using a macro- and a holistic perspective. It provides valuable insights into optimizing and controlling the performance of systems. Chemical engineering systems are characterized by uncertainty arising from poor knowledge of processes and disturbances in systems. This makes optimizing and controlling their behavior a challenge.

The four chapters cover a broad spectrum of topics. While they have been written by researchers working in the areas for several years, the emphasis on each chapter has been on lucidity to enable the graduate student beginning his/her career to develop an interest in the subject. The motivation has been to explain things clearly and at the same time introduce him/her to cutting-edge research in the subject so that the student's interest can be kindled and he/she can feel confident of pursuing a research career in that area.

Chapter 1, by Francois and Bonvin, presents recent developments in the field of process optimization. One of the challenges in systems engineering is an incomplete knowledge of the system. This results in the model of the system being different from that of the plant which it should emulate. In the presence of process disturbances or plant-model mismatch, the classical optimization techniques may not be applicable since they may violate constraints. One way to overcome this is to be conservative. However, this can result in a suboptimal performance. This problem of constraint violation can be eliminated by using information from process measurements. Different methods of measurement-based optimization techniques are discussed in the chapter. The principles of using measurement for optimization are applied to four different problems. These are solved using some of the proposed real-time optimization schemes.

Mathematical models of systems can be developed based on purely statistical techniques. These usually involve a large number of parameters which are estimated using regression techniques. However, this approach does not capture the physics of the process. Hence, its extensions to different conditions may result in inaccurate predictions. This problem is also true of many physical models which contain parameters whose estimates are unknown. These multiparameter estimation problems are not only computationally intensive but may also yield solutions which are physically not realistic. Chapter 2, by Mhamdi and Marquardt, discusses a novel technique of a step-by-step process to address this problem. This is based on the physics prevailing in a system and is computationally elegant. Here the complexity of the problem is increased gradually and the information learnt at each step is used in the next step. Applications of this method to examples in pool boiling, falling films, and reaction diffusion systems are discussed in this chapter.

Wavelets have been gaining prominence as a powerful tool for more than three decades now. They have applications in the fields of signal processing, estimation, pattern recognition, and process systems engineering. Wavelets offer a multiscale framework for signal and system analysis. Here the signals are decomposed into components at different resolutions. Standard techniques are then applied to each of these components. In the area of process systems engineering, wavelets are used for signal compression, estimation, and system identification. Chapter 3, by Tangirala et al., aims to provide an introduction of wavelet transforms to the engineer using an informal approach. It discusses applications in controller loop performance monitoring and multiscale identification. The above are discussed with examples and case studies. It will be very useful to graduate students and researchers in the areas of multiresolution signal processing and also in systems theory and modeling.

In several problems, the need for optimizing more than one objective function simultaneously arises. A typical characteristic could be to define these criteria by using weighting functions and combining the different objective functions into a single objective function. However, a more apt approach is to treat the different objective functions as elements of a vector and determine the optimal solution. Genetic algorithms (GAs) constitute an evolutionary optimization technique. Chapter 4, by Gupta and Garg, discusses the applications of GA to several chemical engineering problems. These applications include industrial reactors and heat exchangers. One of the drawbacks of GA is that it is computationally intensive and hence is slow. This chapter highlights certain modifications of the algorithm which overcomes this limitation of GA. The biomimetic origin of these adaptations provides an interesting avenue for researchers to develop further modifications of GA.

All the above contributions have a heavy dose of mathematics and show different perspectives to address similar problems.

Personally and professionally, it has been a great pleasure for me to be working with all the authors and the editorial team of Elsevier.

S. Pushpavanam

CHAPTER ONE

Measurement-Based Real-Time Optimization of Chemical Processes

Grégory Francois, Dominique Bonvin
Laboratoire d'Automatique, Ecole Polytechnique Fédérale de Lausanne, EPFL, Lausanne, Switzerland

Contents

Advances in Chemical Engineering, Volume 43
ISSN 0065-2377
http://dx.doi.org/10.1016/B978-0-12-396524-0.00001-5

Abstract

This chapter presents recent developments in the field of process optimization. In the presence of uncertainty in the form of plant-model mismatch and process disturbances, the standard model-based optimization techniques might not achieve optimality for the real process or, worse, they might violate some of the process constraints. To avoid constraints violations, a potentially large amount of conservatism is generally introduced, thus leading to suboptimal performance. Fortunately, process measurements can be used to reduce this suboptimality, while guaranteeing satisfaction of process constraints. Measurement-based optimization schemes can be classified depending on the way measurements are used to compensate the effect of uncertainty. Three classes of measurement-based real-time optimization (RTO) methods are discussed and compared. Finally, four representative application problems are presented and solved using some of the proposed RTO schemes.

1. INTRODUCTION

Process optimization is the method of choice for improving the performance of chemical processes while enforcing the satisfaction of operating constraints. Long considered as an appealing tool but only applicable to academic problems, optimization has now become a viable technology (Boyd and Vandenberghe, 2004; Rotava and Zanin, 2005). Still, one of the strengths of optimization, that is, its inherent mathematical rigor, can also be perceived as a weakness, as it is sometimes difficult to find an appropriate mathematical formulation to solve one's specific problem. Furthermore, even when process models are available, the presence of plant-model mismatch and process disturbances makes the direct use of model-based optimal inputs hazardous.

In the past 20 years, the field of "measurement-based optimization" (MBO) has emerged to help overcome the aforementioned modeling difficulties. MBO integrates several methods and tools from sensing technology and control theory into the optimization framework. This way, process optimization does not rely exclusively on the (possibly inaccurate) process model but also on process information stemming from measurements. The first widely available MBO approach was the *two-step approach* that adapts the model parameters on the basis of the deviations between predicted and measured outputs, and uses the updated process model to recompute the optimal inputs (Marlin and Hrymak, 1997; Zhang et al., 2002). Though this approach has become a standard in industry, it has recently been shown that, in the presence of plant-model mismatch, this method is very unlikely to drive the process to optimality (Chachuat et al., 2009). More recently, alternatives to the two-step approach were developed. The *modifier approach* (Marchetti et al., 2009) also proposes to solve a model-based optimization problem but using a fixed plant model. Correction for uncertainty is made via the addition of modifier terms to the cost and the constraint functions of the optimization problem. As the modifiers include information on the deviations between the predicted and the plant necessary conditions of optimality (NCOs), this approach is prone to reach the process optimum upon convergence. Another field has also emerged, for which numerical optimization is not used on-line. With the so-called *self-optimizing approaches* (Ariyur and Krstic, 2003; François et al., 2005; Skogestad, 2000; Srinivasan and Bonvin, 2007), the optimization problem is recast as a control problem that uses measurements to enforce certain optimality features of the real plant.

This chapter reviews these three classes of MBO techniques for both steady-state and dynamic optimization problems. The techniques are motivated and illustrated by four industrial problems that can be addressed via process optimization: (i) the scale-up of optimal operation from the laboratory to production, (ii) the steady-state optimization of continuous production, (iii) the optimal transition between grades in the production of polymers, and (iv) the dynamic optimization of repeated batch processes.

The chapter is organized as follows. The need for improved operation in the chemical industry is addressed, together with the presentation of four application problems. The next section discusses the features of chemical processes that are relevant to optimization. Then, the basic elements of static and dynamic optimization are presented, followed by an in-depth exposure of MBO and the three aforementioned classes of techniques. Then, the four case studies are presented, followed by conclusions.

2. IMPROVED OPERATION OF CHEMICAL PROCESSES

2.1. Need for improved operation in chemical production

In a world of growing competition, every tool or method that leads to the reduction of production costs or the increase of benefits is valuable. From this point of view, the chemical industry is no different. As a consequence of this increasing competition, the structure of the chemical industry has progressively moved from the manufacturing of basic chemicals to a much more segmented market including basic chemicals, life sciences, specialty chemicals and consumer products (Choudary et al., 2000). This segmentation in terms of the nature of the products impacts the structural organization of the companies (Bonvin et al., 2006), the interaction between the suppliers and the customers, but also, on the process engineering side, the nature and the capacity of the production units, as well as the criterion for assessing the production performance. This segmentation is briefly described next.

1. "Basic chemicals" are generally produced by large companies and sold to a large number of customers. As profit is generally ensured by the high-volume production (small margins but propagated over a large production), one key for competitiveness lies in the ability of following the market fluctuations so as to produce the right product, at the right quality, at the right instant. Basic chemicals, also referred to as "commodities," encompass a wide range a products or intermediates such as monomers, large-volume polymers (PE, polyethylene; PS, polystyrene; PP, polypropylene; PVC, polyvinyl chloride; etc), inorganic chemically (salt, chlorine, caustic soda, etc.) or fertilizers.
2. Active compounds used in consumer goods and industrial products are referred to as "fine chemicals." The objective of fine-chemicals companies is typically to achieve the required qualities of the products, as given by the customers (Bonvin et al., 2001). Hence, the key to being competitive is generally to provide the same quality as the competitors at a lower price or to propose a higher quality at a lower or equal price. Examples of fine chemicals include advanced intermediates, drugs, pesticides, active ingredients, vitamins, flavors, and fragrances.
3. "Performance chemicals" correspond to the family of compounds, which are produced to achieve well-defined requirements. Adhesives, electrochemicals, food additives, mining chemicals, pharmaceuticals, specialty polymers, and water treatment chemicals are good representatives of this class of products. As the name implies, these chemicals are critical to the performance of the end products in which they are used. Here, the competitiveness of performance-chemicals companies relies highly on their ability to achieve these requirements.
4. Since "specialty chemicals" encompass a wide range of products, this segment consists of a large number of small companies, more so than other segments of the chemical industry (Bonvin et al., 2001). In fact, many specialty chemicals are based on a single product line, for which the company has developed a leading technology position.

While basic chemicals are typically produced at high volumes in continuous operation, fine chemicals, performance chemicals and specialty chemicals are more widely produced in batch reactors, that is, low-volume, discontinuous production. However, regardless of the type of chemicals that are produced or the nature and size of the production units, in such a competitive industry sector, it is of paramount importance to optimize key business drivers such as product quality and production efficiency to maintain a competitive advantage in a global market weighing more than 1.6 trillion USD per year.

2.2. Four representative application challenges

In this section, we describe four typical challenges that the chemical industry has to deal with for improving production. We also show that, although they appear to be different in nature, these problems can be formulated in a very similar manner and solved with well-chosen optimization techniques.

2.2.1 Scaling up reactor operation from lab size to plant size

This problem is very common in industry. Suppose that a promising route for producing some new high-value-added chemical has been investigated. Laboratory experiments provide either a set of constant operating conditions for the case of a continuous stirred-tank reactor (CSTR), or input profiles for batch or fed-batch reactors. The resulting recipe is generally appropriate from a chemical viewpoint, as the chemists in charge of process development have optimized various factors such as temperature, pressure, concentration, and feed rates. However, this optimality property only holds for the reactor or the experimental facility it has been designed for, and it is very unlikely that these conditions will also be optimal for production in large reactors. For example, the mixing and heat-transfer characteristics in a 10-ton production reactor are quite different from those found in a 1-L laboratory reactor. Hence, it is necessary to adjust these conditions, with the main questions being which variables to adjust and how. One solution would be to use pilot-plant investigation, on a mid-size reactor, to fill the significant gap between the laboratory and the production scales. However, and this is particularly true for small companies, the trend today is to jump over the pilot-plant investigations by using systematic techniques for scaling up the process. We will see thereafter that run-to-run optimization methods are well suited to meet this challenging goal.

2.2.2 Steady-state optimization of continuous operation

Consider the continuous production of some chemicals, for which optimal performance is achieved when all units operate around optimal, yet unknown, set points. The determination of these set-point values is, by itself, already a difficult issue that can be solved using model-based optimization, provided a model is available. However, because of market fluctuations as well as variations of the demand and of the raw materials and energy costs, the optimal operating conditions are very likely to vary with time. Hence, these model-based optimal operating conditions need to be adjusted in real-time to maintain optimality.

This challenge is illustrated by means of the optimization of a solid oxide fuel cell stack, a system that needs to be operated at maximal electrical efficiency to be cost effective. In addition, the stack should always be able to track the load changes, that is, produce the power required by the users. In our fuel cell example, drastic changes in the power demand call for fast and reliable adaptation of the operating conditions. As the exogenous changes and perturbations in a large chemical production unit are much slower, the adaptation of the operating conditions need not be fast. Hence, the fuel cell example can be seen as a fast version of what would occur in a large chemical production unit. Yet, the goal is the same, namely, to be able to adjust the operating conditions at more or less the speed of the demand changes.

2.2.3 Optimal grade transition

The third case study deals with a very frequent industrial challenge. Consider a continuous stirred-tank reactor operated at steady state to manufacture product A. As seen in the previous problem, the operating conditions need to be adjusted in real-time to respond to market fluctuations. However, it may happen that market fluctuations or customer orders require to move to the production of another product, referred to as B, whose formulation is sufficiently close to A so that there is no need to stop production. The operating conditions have to be adjusted to bring the reactor at the optimal operating conditions for B. In practice, it is often desired to perform this transition in an optimal manner as, between two grades, raw materials and energy are being consumed and the workforce is still around, while generally no useful product is produced. When grade transitions are frequent, this can lead to significant losses, and minimizing the duration of the transient as well as the raw materials/product losses become clear objectives. The example thereafter will address the optimization of the grade transition in polyethylene reactors.

2.2.4 Run-to-run optimization of batch polymerization processes

The fourth problem concerns the optimization of batch processes. A batch (or semi-batch) process exhibits no steady state. Reactants are placed into the reactor before the reaction starts; semi-batch processes also include the addition of some of the reactants during the reaction. When the reaction is thought to be finished, the operation is stopped, the reactor is opened and the products recovered. The typical challenge is to determine the control policy, that is, the feeding and temperature profiles that optimize some performance criterion (such as yield, conversion, purity, reaction time, energy consumption), while guaranteeing the satisfaction of both operational constraints during the batch as well as quality and production constraints at final time. Model-based optimization techniques can be used for this purpose.

Another particularity of batch processes lies in their repetitive nature, which opens up the possibility to iteratively improve performance by using past data to adjust the input profiles for future batches. In practice, the adjustments are often guided by experience. We will consider the run-to-run optimization of an industrial emulsion copolymerization reactor to show how adjustments can be performed in a systematic manner.

3. OPTIMIZATION-RELEVANT FEATURES OF CHEMICAL PROCESSES

3.1. Presence of uncertainty

In practice, the presence of uncertainty makes process improvement difficult. Uncertainty is a vague notion as it incorporates everything that is not known with certainty such as structural plant-model mismatch, parametric errors and process disturbances. This definition of uncertainty assumes that a plant model is available, that is, a set of differential and algebraic equations that mimic the plant behavior. Plant-model mismatch incorporates all the structural differences between the plant and its model such as neglected dynamics and simplified nonlinearities, while parametric errors deal with the fact that some of the model parameters are not known accurately. In addition, there are process disturbances.

As shown in Fig. 1.1, process disturbances enter at all levels of the process control architecture. Slow disturbances like market fluctuations will typically impact the decisions taken at the planning and scheduling level, while fast disturbances such as pressure variations are typically dealt with at the process control level. The optimization layer faces medium-term disturbances such as catalyst decay and changes in raw material quality.

Similarly to what is performed in the control layer, where measurements are compared to set points to compute control actions that ensure set-point tracking, measurements can also be used in the upper two layers. More specifically, the optimization layer incorporates information from both the

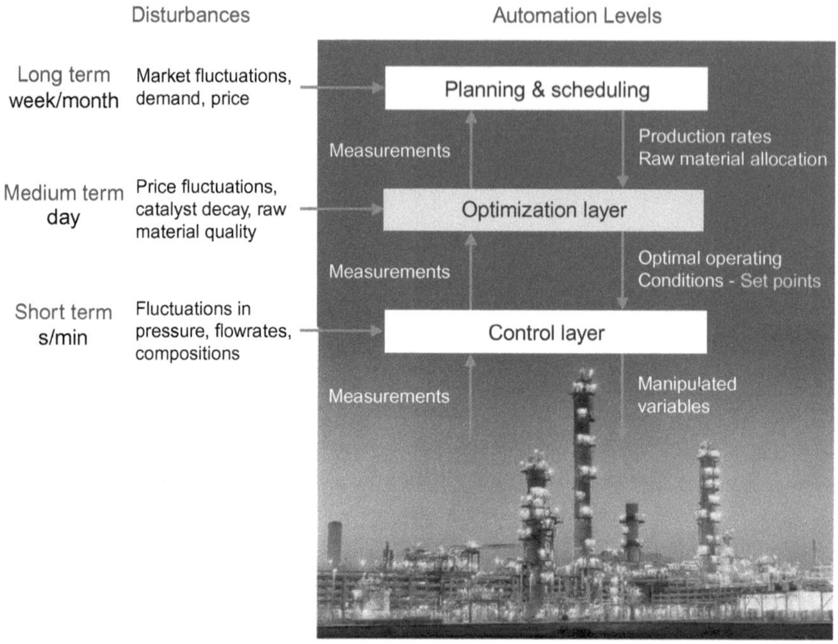

Figure 1.1 Disturbances affecting the various levels of process automation. (For color version of this figure, the reader is referred to the online version of this chapter.)

control and planning layers to update the set points of the low-level controllers, thereby rejecting the effect of medium-term disturbances. This gives rise to the framework of MBO, which will be detailed in the forthcoming sections.

3.2. Presence of constraints

Process improvement is also affected by the presence of constraints, which are incorporated in the optimization problem. The constraints include input bounds, which correspond to the saturation of actuators (e.g., maximal opening of a valve, maximal flow rate of a pump, minimal cooling fluid temperature) as well as limits on some state and output variables. The satisfaction of process constraints ensures that the process is operated safely and the products meet prespecified requirements. However, as optimizing a process amounts to pushing it to its limits, the optimal solution often turns out to be on some of the constraints. Model uncertainty is therefore very detrimental, as the model-based optimal solution may violate plant constraints. In fact, in many applications, it is often preferred to be suboptimal if it means that the constraints are more likely to be satisfied. One solution is to monitor and track the constraints. Tracking the active constraints, that is, keeping these constraints active despite uncertainty, can be a very effective way of implementing an optimal policy. When the set of active constraints fully determines the optimal inputs, provided this set does not change with uncertainty, constraint tracking is indeed optimal.

3.3. Continuous versus batch operation

Another feature that affects both the formulation and the solution of the optimization problem is the nature of the operation. As seen before, processes can be divided into two categories, namely, steady-state and transient processes. Transient processes are characterized by the presence of initial and terminal conditions and the absence of a steady state. In a transient process, the optimal solution indicates how to drive the process from its initial to its terminal state in some optimal way. For this purpose, the optimization problem is formulated as a dynamic optimization problem. In contrast, the optimization of a steady-state process calls for static optimization. However, as will be seen later, transient information can also be used for determining optimal steady-state conditions.

3.4. Repetitive nature of batch processes

Finally, transient processes, such as batch or semi-batch processes, are generally repeated over time. This repetitive nature can be exploited to implement run-to-run (or batch-to-batch) optimization. The key feature is the use of measurements from past batches to update the control policy of future batches, again with the objective of improving performance and enforcing the satisfaction of active constraints.

4. MODEL-BASED OPTIMIZATION

Apart from very specific cases, the standard way of solving an optimization problem is via numerical optimization. For this purpose, a model of the process is required. A steady-state model leads to a static optimization problem (or nonlinear program, NLP) with a finite number of time-invariant decision variables, whereas a dynamic model calls for the determination of a vector of input profiles via dynamic optimization.

4.1. Static optimization and KKT conditions

4.1.1 Problem formulation

Consider the following steady-state constrained optimization problem:

$$\begin{aligned} \min_{\mathbf{u}} \;& J := \varphi(\mathbf{u},\mathbf{y}) \\ \text{s.t.} \;& \mathbf{h}(\mathbf{u},\mathbf{y}) = \mathbf{0} \\ & \mathbf{g}(\mathbf{u},\mathbf{y}) \le \mathbf{0} \end{aligned} \qquad [1.1]$$

where J is the scalar cost to be minimized, $\mathbf{y}$ the n_y-dimensional output vector, $\mathbf{u}$ the m-dimensional vector of time-invariant inputs, $\mathbf{g}$ the n_g-dimensional vector of constraints, and $\mathbf{h}(\mathbf{u},\mathbf{y})$ the steady-state model linking input and ouput variables. With this formulation, the vector of constraints can include pure input, pure output or mixed input-output constraints.

Provided the outputs can be expressed explicitly in terms of the inputs, that is, $\mathbf{y} = \mathbf{H}(\mathbf{u})$, the steady-state optimization problem can be reformulated as follows:

$$\begin{aligned} \min_{\mathbf{u}} \;& J = \varphi(\mathbf{u},\mathbf{H}(\mathbf{u})) \\ \text{s.t.} \;& \mathbf{g}(\mathbf{u},\mathbf{H}(\mathbf{u})) \le \mathbf{0} \end{aligned} \qquad [1.2]$$

or equivalently

$$\begin{aligned} \min_{\mathbf{u}} \;& J = \Phi(\mathbf{u}) \\ \text{s.t.} \;& \mathbf{G}(\mathbf{u}) \le \mathbf{0} \end{aligned} \qquad [1.3]$$

4.1.2 KKT necessary conditions of optimality

With the formulation (1.3) and the assumption that the cost and constraint functions are differentiable, the Karush–Kuhn–Tucker (KKT) conditions read (Bazarra et al., 1993):

$$\begin{aligned} \mathbf{G}(\mathbf{u}^*) &\le \mathbf{0} \\ \nabla\Phi(\mathbf{u}^*) + \boldsymbol{\nu}^{*T}\nabla\mathbf{G}(\mathbf{u}^*) &= \mathbf{0} \\ \boldsymbol{\nu}^* &\ge \mathbf{0} \\ \boldsymbol{\nu}^{*T}\mathbf{G}(\mathbf{u}^*) &= 0 \end{aligned} \qquad [1.4]$$

where $\mathbf{u}^*$ denotes the candidate solution, $\boldsymbol{\nu}^*$ the n_g-dimensional vector of Lagrange multipliers associated with the constraints, $\nabla\Phi(\mathbf{u}^*)$ the m-dimensional row vector denoting the cost gradient evaluated at $\mathbf{u}^*$, and $\nabla\mathbf{G}(\mathbf{u}^*)$ the $(n_g \times m)$-dimensional Jacobian matrix computed at $\mathbf{u}^*$. For these equations to be necessary conditions, $\mathbf{u}^*$ needs to be a regular point for the constraints, which calls for linear independence of the active constraints, that is, $\text{rank}\{\nabla\mathbf{G}_a(\mathbf{u}^*)\} = n_{g,a}$, where $\mathbf{G}_a$ represents the set of active constraints, whose cardinality is $n_{g,a}$.

The first condition in Eq. (1.4) is referred to as the primal feasibility condition, while the fourth one is called the complementarity slackness condition; the second and third conditions are called the dual feasibility conditions. The second condition indicates that, at the optimal solution, collinearity between the cost gradient and the constraint gradient prevents from finding a search direction that would result in cost reduction while still keeping the constraints satisfied.

4.1.3 Solution methods

Static optimization can be solved by state-of-the-art nonlinear programming techniques. In the presence of constraints, the three most popular approaches are (Gill et al., 1981): (i) penalty function methods, (ii) interior-point methods, and (iii) sequential quadratic programming (SQP).

The main idea in penalty function methods is to replace the solution of a constrained optimization problem by the solution of a sequence of unconstrained optimization problems. This is made possible by incorporating the constraints in the objective function via a penalty term, which penalizes any violation of the constraints while guaranteeing that the two problems share the same solution (by selecting weighting coefficients that are sufficiently large).

Interior-point methods also incorporate the constraints in the objective function (Forsgren et al., 2002). The constraints are approached from the feasible region, and the additive terms increase to become infinitely large at the value of the constraints, thereby acting more like a barrier than a penalty term. A clear advantage of interior-point methods is that feasible iterates are generated, while for penalty function methods, feasibility is only ensured upon convergence. Note that Srinivasan et al. (2008) have proposed a barrier-penalty function that combined the advantages of both approaches.

Another way of computing the solution of a static optimization problem is to find a solution to the set of NCOs, for example using SQP iteratively. SQP methods solve a sequence of optimization subproblems, each one minimizing a quadratic approximation to the Lagrangian function $L = \Phi + \boldsymbol{\nu}^T \mathbf{G}$ subject to a linear approximation of the constraints. SQP typically uses Newton's or quasi-Newton methods to solve the KKT conditions (Gill et al., 1981).

4.2. Dynamic optimization and PMP conditions

4.2.1 Problem formulation

Consider the following constrained dynamic optimization problem:

$$\begin{aligned}
\min_{\mathbf{u}(t),\boldsymbol{\rho}} \quad & J := \varphi(\mathbf{x}(t_\mathrm{f}),\boldsymbol{\rho}) \\
\text{s.t.} \quad & \dot{\mathbf{x}} = \mathbf{F}(\mathbf{u}(t),\mathbf{x}(t),\boldsymbol{\rho}) \quad \mathbf{x}(0) = \mathbf{x}_0 \\
& \mathbf{S}(\mathbf{u}(t),\mathbf{x}(t),\boldsymbol{\rho}) \leq \mathbf{0} \\
& \mathbf{T}(\mathbf{x}(t_\mathrm{f}),\boldsymbol{\rho}) \leq \mathbf{0}
\end{aligned} \qquad [1.5]$$

where φ is the terminal-time cost functional to be minimized, $\mathbf{x}(t)$ the n-dimensional vector of states profiles with the known initial conditions $\mathbf{x}_0$, $\mathbf{u}(t)$ the m-dimensional vector of input profiles, $\boldsymbol{\rho}$ the n_ρ-dimensional vector of time-invariant decision variables, $\mathbf{S}$ the n_S-dimensional vector of path constraints, $\mathbf{T}$ the n_T-dimensional vector of terminal constraints, and t_f the final time, which can be either free or fixed. If t_f is free, it is part of $\boldsymbol{\rho}$. The optimization problem (Eq. 1.5) is said to be in the *Mayer* form, that is, J is a terminal-time cost functional. When an integral cost is added to φ, the corresponding problem is said to be in the *Bolza* form, while when it only incorporates the integral cost, it is referred to as being in the *Lagrange* form. However, it is straightforward to show that these three formulations are equivalent by the introduction of additional states.

4.2.2 Pontryagin's minimum principle

The NCOs for a dynamic optimization problem are given by Pontryagin's minimum principle (PMP). Although less tractable and more difficult to interpret than the KKT conditions, application of PMP can provide the same insight by separating active and inactive constraints. Upon defining:

- the Hamiltonian function

$$H(t) = \boldsymbol{\lambda}^T(t)\mathbf{F}(\mathbf{u}(t),\mathbf{x}(t),\boldsymbol{\rho}) + \boldsymbol{\mu}^T(t)\mathbf{S}(\mathbf{u}(t),\mathbf{x}(t),\boldsymbol{\rho})$$

  and the augmented terminal cost

$$\Phi(t_\mathrm{f}) = \varphi(\mathbf{x}(t_\mathrm{f}),\boldsymbol{\rho}) + \boldsymbol{\nu}^T\mathbf{T}(\mathbf{x}(t_\mathrm{f}),\boldsymbol{\rho})$$

  where $\boldsymbol{\lambda}(t)$ are the adjoint variables such that

$$\dot{\boldsymbol{\lambda}}^T(t) = -\frac{\partial H}{\partial \mathbf{x}}(t), \quad \boldsymbol{\lambda}^T(t_\mathrm{f}) = \frac{\partial \Phi}{\partial \mathbf{x}}(t_\mathrm{f}),$$

  $\boldsymbol{\mu}(t) \geq \mathbf{0}$ are the Lagrange multipliers associated with the path constraints, and $\boldsymbol{\nu} \geq \mathbf{0}$ are the Lagrange multipliers associated with the terminal constraints,
- the total terminal cost $\Psi(t_\mathrm{f}) = \Phi(t_\mathrm{f}) + \int_0^{t_\mathrm{f}} H(t)\mathrm{d}t$, the NCOs can be expressed as given in Table 1.1 (Srinivasan et al., 2003).

Table 1.1 NCOs for a dynamic optimization problem

	Path	Terminal
Constraints	$\boldsymbol{\mu}^T\mathbf{S}=0, \quad \boldsymbol{\mu}\geq 0$	$\boldsymbol{\nu}^T\mathbf{T}=0, \quad \boldsymbol{\nu}\geq 0$
Sensitivities	$\frac{\partial H}{\partial \mathbf{u}}=\mathbf{0}$	$\frac{\partial \Psi}{\partial \boldsymbol{\rho}}=\mathbf{0}$

The solution obtained will generally be discontinuous and consist of several intervals or arcs. Each interval will be characterized by a different set of active path constraints, that is, this set changes between successive intervals.

4.2.3 Solution method

Solving the dynamic optimization problem of Eq. (1.5) corresponds to finding the best optimal control profiles $\mathbf{u}(t)$ and the best time-invariant decision variables $\boldsymbol{\rho}$ such that the cost functional is minimized, while meeting both the path and terminal constraints. As the decision variables $\mathbf{u}(t)$ are infinite dimensional, the inputs need to be parameterized using a finite set of parameters in order to utilize numerical techniques. These techniques are classified into two main categories according to the underlying formulation, namely, the direct optimization methods that solve the optimization problem (Eq. 1.5) directly, and the PMP-based methods that attempt to satisfy the NCOs given in Table 1.1.

Direct optimization methods are distinguished further depending on whether the system equations are integrated explicitly or not. In the *sequential approach*, the system equations are integrated explicitly, and the optimization is carried out in the space of the input variables only. This corresponds to a "feasible" path approach as the differential equations are satisfied at each step of the optimization. A piecewise-constant or piecewise-polynomial approximation of the inputs is often used. The most computationally intensive part of the sequential approach is the accurate integration of the system equations, even when the decision variables are far from the optimal solution. In the *simultaneous approach*, an approximation of the system equations is introduced to avoid explicit integration for each candidate input profile, thereby reducing the computational burden. As the optimization is carried out in the full space of discretized inputs and states, the differential equations are satisfied only at the solution of the optimization problem (Vassiliadis et al., 1994). This is therefore called an "infeasible path" approach. The direct approaches are by far the most commonly used. Note, however, that input parameterization is often chosen arbitrarily by the user, which can affect the efficiency and the accuracy of the approach.

PMP-based methods try to satisfy the first-order NCOs given in Table 1.1. The NCOs involve the state and adjoint variables, which need to be computed via integration. The differential equation system is a two-point boundary value problem as initial conditions are available for the states and terminal conditions for the adjoints. The optimal inputs can be expressed analytically in terms of the states and the adjoints from the NCOs, that is, $\mathbf{u}^* = \mathrm{U}(\mathbf{x}, \boldsymbol{\lambda})$. The resulting differential-algebraic system of equations can be solved using a shooting approach (Bryson, 1999), that is, the decision variables include the initial conditions $\boldsymbol{\lambda}(0)$ that are chosen in order to satisfy $\boldsymbol{\lambda}(t_f)$.

4.3. Effect of plant-model mismatch

4.3.1 Plant-model mismatch

The model used for optimization consists of a set of equations that represent an abstract view, yet always a simplification of the real process. Such a model is built based on conservations laws (mass, numbers of moles, energy) and constitutive relationships to express kinetics, equilibria and transport phenomena. The simplifications that are introduced at the modeling stage to obtain a tractable model affect the quality of the process model in two ways: (i) some physical or chemical phenomena are ignored or assumed to be negligible, and (ii) some dynamic equations are assumed to be at quasi-steady state or are simply removed for the sake of simplicity. Hence, the structure of the working model invariably differs from that of the idealized "true model." This is the so-called structural plant-model mismatch, which affects the quality of model predictions. The resulting model involves a number of physical parameters, whose values are not known accurately. These parameters are identified using process measurements and, consequently, are only known to belong to some confidence intervals with a certain probability. For the sake of simplicity, we will consider thereafter that all modeling uncertainties, though unknown, are incorporated in the vector of uncertain parameters $\boldsymbol{\theta}$.

4.3.2 Model adequacy

Uncertainty is detrimental to the quality of both model predictions and optimal solutions. If the model is not able to predict the process outputs accurately, it will most likely not be able to predict correctly its NCOs. On the other hand, even if the model is able to predict the process outputs accurately, it will often be unable to predict the NCOs correctly as it has been trained to predict the outputs and not, for instance, the cost and constraint gradients. Hence, if model-based optimization techniques are successful in computing optimal inputs for the model, they typically fail to find those for the plant. The effect of plant-model mismatch can be visualized by writing down the corresponding optimization problems for the model and the plant, here for the steady-state case:

$$\underbrace{\begin{array}{ll}\min\limits_{\mathbf{u}} & J_{\mathrm{p}} = \Phi_{\mathrm{p}}(\mathbf{u}) := \varphi(\mathbf{u}, \mathbf{y}_{\mathrm{p}}) \\ \text{s.t.} & \mathbf{y}_{\mathrm{p}} = \mathbf{H}_{\mathrm{p}}(\mathbf{u}) \\ & \mathbf{G}_{\mathrm{p}}(\mathbf{u}) = \mathbf{g}(\mathbf{u}, \mathbf{y}_{\mathrm{p}}) \leq \mathbf{0}\end{array}}_{\text{plant optimization}} \qquad \underbrace{\begin{array}{ll}\min\limits_{\mathbf{u}} & J = \Phi(\mathbf{u}) \\ \text{s.t.} & \mathbf{G}(\mathbf{u}) \leq \mathbf{0}\end{array}}_{\text{model optimization}} \qquad [1.6]$$

where $\mathbf{y}_{\mathrm{p}}$ is the n_y-dimensional vector of plant outputs, with the subscript $(.)_{\mathrm{p}}$ denoting the plant. The plant is seen as the mapping $\mathbf{y}_{\mathrm{p}} = \mathbf{H}_{\mathrm{p}}(\mathbf{u})$ of the manipulated inputs to the measured outputs. As these two optimization problems are different, their NCOs are different as well. The property that ensures that a model-based optimization problem will be able to determine the optimal inputs for the plant is referred to in the literature as "model adequacy." A model is adequate if and only if it generates the solution $\mathbf{u}^*$ that satisfies the plant NCOs, that is:

$$\begin{aligned} \mathbf{G}_{\mathrm{p}}(\mathbf{u}^*) &\leq \mathbf{0} \\ \nabla \Phi_{\mathrm{p}}(\mathbf{u}^*) + \boldsymbol{\nu}_{\mathrm{p}}^{*T} \nabla \mathbf{G}_{\mathrm{p}}(\mathbf{u}^*) &= \mathbf{0} \\ \boldsymbol{\nu}_{\mathrm{p}}^* &\geq \mathbf{0} \\ \boldsymbol{\nu}_{\mathrm{p}}^{*T} \mathbf{G}_{\mathrm{p}}(\mathbf{u}^*) &= 0 \end{aligned} \qquad [1.7]$$

In other words, the model should be able to predict the correct set of active plant constraints (rather than model constraints) and the correct alignment of plant gradients (rather than model gradients). Model adequacy represents a major challenge in process optimization as, as discussed earlier, models are trained to predict the plant outputs rather than the NCOs. In practice, application of the model-based optimal inputs leads to suboptimal, and often infeasible, operation.

5. MEASUREMENT-BASED OPTIMIZATION

One way to reject the effect of uncertainty on the overall performance (optimality and feasibility) is by adequately incorporating process measurements in the optimization framework. In fact, this is exactly how controllers work. A controller is typically designed and tuned using a process model. If the model is an exact copy of the plant to control, the controller performance will be exactly the same as with model-based simulation. Although this is never the case, the controller still performs well in terms of set-point tracking and disturbance rejection. This robustness to modeling errors is provided by the feedback of process measurements, with the control action using only the difference between measurements and set points. MBO schemes exhibit the same features, that is, ensure optimality despite modeling errors through appropriate feedback.

5.1. Classification of measurement-based optimization schemes

Measurements can be incorporated in different ways in the optimization framework. This section aims at classifying MBO schemes according to the way measurements are used and feedback is implemented. Real-time optimization (RTO) corresponds to the "optimization layer" in Fig. 1.1. Its main objective is to process measurements from the plant to compute optimal set points (inputs) for the low-level controllers so as to track the plant NCOs.

Real-time input adaptation is required because uncertainty can change the optimal operating conditions. We consider next three ways of modifying these inputs: (i) adapt the process model that is used subsequently for optimization, (ii) adapt the optimization problem and repeat the optimization, and (iii) directly adapt the inputs through an appropriate feedback strategy. The two former are explicit optimization techniques as the optimization problem is solved numerically (along the line of direct optimization methods), while the latter is an implicit scheme as optimality and feasibility are enforced via feedback control rather than numerical optimization (along the line of PMP-based methods). These three MBO schemes are shown in Fig. 1.2.

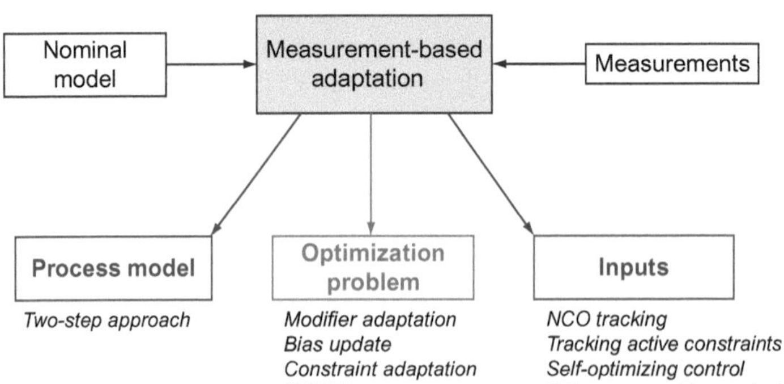

Figure 1.2 Classification of measurement-based optimization schemes (ISOPE stands for "integrated system optimization and parameter estimation"). (For color version of this figure, the reader is referred to the online version of this chapter.)

5.2. Implementation aspects

MBO techniques also differ in the way measurements are used. Some of the methods only use the current on-line measurements, while other methods also incorporate past data. This is of course closely related to the nature of the process at hand. For instance, batch processes, which are repeated over time, are natural candidates for incorporating past data. Four MBO implementation types can be distinguished based on the nature of the control (on-line or run-to-run) and the objectives (run-time or run-end):

5.2.1 On-line control of run-time objectives

This control strategy can be applied to both continuous and discontinuous processes. For example, when the optimal strategy calls for tracking the active constraints $\mathbf{y}_{\text{ref}}(t)$, this can be performed with simple on-line controllers that keep the controlled constraints active. Optimality can be ensured this way when the number of active constraints equals the number of inputs. The control laws can be written generically as:

$$\mathbf{u}_k(t) = \kappa\left(\mathbf{y}_{\text{p},k}(t), \mathbf{y}_{\text{ref}}(t)\right) \quad [1.8]$$

where the subscript k, which denotes the kth batch in the case of batch processes, is simply removed in the case of continuous operation.

5.2.2 On-line control of run-end outputs

The idea here is to use on-line measurements to control run-end outputs. An example is the control of an active terminal constraint in a batch process. The standard way of implementing such a control policy is to use on-line measurements combined with model-based prediction of the terminal constraint via, for example, model predictive control (MPC). The controller can be written generically as:

$$\mathbf{u}_k(t) = \kappa\left(\mathbf{y}_{\text{pred},k}(t), \mathbf{y}_{\text{ref}}\right) \quad [1.9]$$

where $\mathbf{y}_{\text{pred}}(t)$ and $\mathbf{y}_{\text{ref}}$ denote the prediction at time t of terminal quantities and the corresponding run-end set points, respectively.

5.2.3 Run-to-run control of run-time outputs

In contrast to the two aforementioned strategies, for which the control action is computed at every sampling instant, the idea here is to control run-time outputs by taking decisions at a slower time scale. Iterative learning control (ILC) is a good example of such control, as decisions are taken prior to a run to control run-time outputs (Moore, 1993). Clearly, this strategy exhibits the limitations of open-loop control for run-time operation, in particular the fact that there is no feedback correction for run-time disturbances. Yet, this scheme is highly efficient for generating feedforward input terms. The controller has the following generic structure:

$$\mathbf{u}_{k+1}[0,t_f] = I\left(\mathbf{y}_{p,k}[0,t_f], \mathbf{y}_{ref}[0,t_f]\right) \qquad [1.10]$$

where $\mathbf{y}_{ref}[0,t_f]$ denotes the desired profiles of the run-time outputs. The ILC controller processes the entire profile of the current run to generate the entire manipulated profile for the next run.

5.2.4 Run-to-run control of run-end objectives

Steady-state optimization of continuous processes and run-to-run optimization of discontinuous processes can be performed in a similar way. For the steady-state optimization of continuous processes, input values are applied to the process at the *k*th iteration and measurements are taken once steady state has been reached. Based on these measurements, an optimization problem is solved to determine the inputs for iteration $k+1$. The run-to-run optimization of discontinuous processes is implemented in a similar manner. Input profiles are applied in an open-loop manner to the *k*th batch. Upon completion of the batch, measurements taken during the batch and at the end of the batch are used for updating the input profiles for batch $k+1$. Upon parameterization of the input profiles using a finite number of parameters, that is, $\mathbf{u}_k[0,t_f] = \mathrm{U}(\boldsymbol{\pi}_k)$, the run-to-run control law can be written generically as:

$$\boldsymbol{\pi}_{k+1} = \Re\left(\mathbf{y}_{p,k}(t_f), \mathbf{y}_{ref}(t_f)\right) \qquad [1.11]$$

where $\mathbf{y}_{ref}(t_f)$ represents the run-end objectives.

5.3. Two-step approach

5.3.1 Basic idea

In the two-step approach, measurements are used to refine the model, and the input update is obtained by solving the optimization problem using the refined model (Marlin and Hrymak, 1997; Zhang et al., 2002). The two-step approach can be applied to both dynamic and steady-state optimization problems. Optimization is performed iteratively, that is, in a run-to-run manner for dynamic processes and from one steady state to the next for continuous processes. The two-step approach has gained popularity over the past 30 years mainly because of its conceptual simplicity. Yet, the two-step

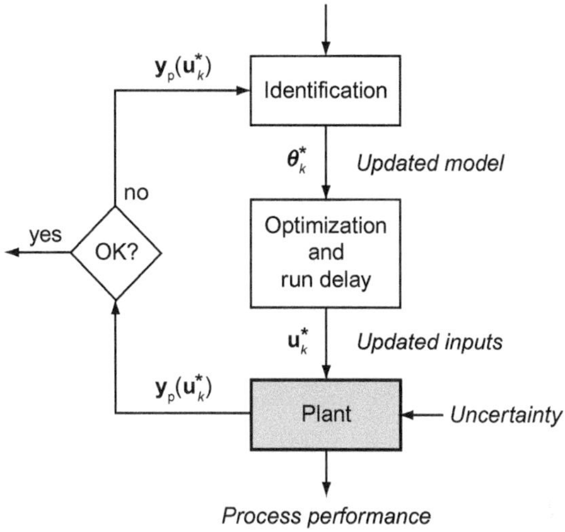

Figure 1.3 Basic idea of the two-step approach. (For color version of this figure, the reader is referred to the online version of this chapter.)

approach is characterized by certain intrinsic difficulties that are often overlooked.

In its iterative version, the two-step approach involves two optimization problems, namely, one each for parameter identification and process optimization (Fig. 1.3). For the static (or steady-state) optimization case, the two problems are as follows:

$$\begin{aligned} \text{Identification:} \quad & \boldsymbol{\theta}_k^* := \underset{\boldsymbol{\theta}}{\operatorname{argmin}} \left(\left\| \mathbf{y}_\text{p}\left(\mathbf{u}_k^*\right) - \mathbf{y}\left(\mathbf{u}_k^*, \boldsymbol{\theta}\right) \right\| \right) \\ & \text{s.t.} \quad \boldsymbol{\theta} \in \Theta \\ \text{Optimization:} \quad & \mathbf{u}_{k+1k}^* := \underset{\mathbf{u}}{\operatorname{argmin}} \, \Phi\left(\mathbf{u}, \boldsymbol{\theta}_k^*\right) \\ & \text{s.t.} \quad \mathbf{G}\left(\mathbf{u}, \boldsymbol{\theta}_k^*\right) \leq \mathbf{0} \end{aligned} \qquad [1.12]$$

where Θ indicates the set in which the uncertain parameters $\boldsymbol{\theta}$ are assumed to lie.

The first step identifies best values for the uncertain parameters by minimizing some norm of the output prediction error. The second step then computes the optimal inputs for the updated model. Algorithmically, the optimization of the steady-state performance of a continuous process proceeds as follows:

i. Apply the model-based optimal inputs to the real process $\mathbf{u}_k^*$.

ii. Wait until steady state is reached and compute the distance between the predicted and measured steady-state outputs.

iii. Continue if this distance exceeds the tolerance, otherwise stop.
iv. Solve the identification problem to obtain $\boldsymbol{\theta}_k^*$.
v. Solve the optimization problem to obtain $\mathbf{u}_{k+1}^*$.
vi. Set $k := k + 1$ and go back to (i).

The two-step approach suffers from two main limitations. First, the identification problem requires sufficient excitation. However, as the inputs are computed for optimality rather than for performing identification, there is often insufficient excitation for the purpose of identification. The second limitation is inherent to the philosophy of the method. The model update is driven by the output prediction error, and the adjustable handles are the model parameters. Hence, the method assumes that (i) all the uncertainty (including process disturbances) can be represented by the set of uncertain parameters $\boldsymbol{\theta}$. Figure 1.3 depicts the philosophy of the two-step approach, where input update results from the adaptation of the model parameters.

5.3.2 Model adequacy

The problem of model selection in the two-step RTO approach has been discussed in Forbes and Marlin (1996). If the model is structurally correct and the parameters are identifiable, convergence to the plant optimum can be achieved in one iteration. However, in the presence of plant-model mismatch, whether the scheme converges, or to which point it does converge, becomes anyone's guess. This is due to the fact that the objective of parameter adaptation might be unrelated to the cost and constraints that drive optimality in the optimization problem. Hence, minimizing the mean-square error of the plant outputs may not help in the quest for feasibility and optimality. Convergence under plant-model mismatch has been addressed by Biegler et al. (1985) and Forbes et al. (1994), where it was shown that optimal operation is reached if model adaptation leads to matched KKT conditions for the model and the plant. We will show next that this is rarely the case in the presence of structural plant-model mismatch, because the two-step approach has typically too few degrees of freedom.

Consider the two-step RTO scheme at the kth iteration, with the estimation and optimization problems given by Eq. (1.12). The top part of Fig. 1.4 illustrates the iterative scheme, whereby the optimization problem uses the best estimate $\boldsymbol{\theta}_k^*$ from the parameter estimation problem to compute the next input $\mathbf{u}_{k+1}^*$. A plant model is *adequate* for optimization if parameter values, say $\bar{\boldsymbol{\theta}}$, can be found such that a fixed point of the RTO scheme coincides with the plant optimum $\mathbf{u}_p^*$. Let us assume that the model is adequate, that is, the iterative scheme has converged to the true plant optimum, with

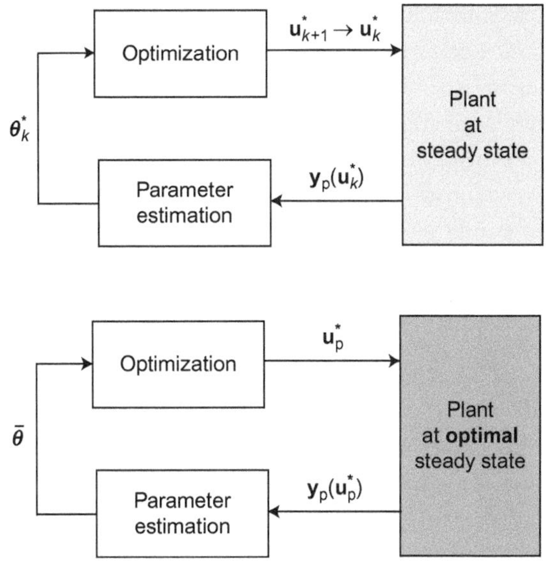

Figure 1.4 Two-step approach with the parameter estimation and the optimization problems. Top: iterative scheme; bottom: ideal situation upon convergence to the plant optimum. (For color version of this figure, the reader is referred to the online version of this chapter.)

the converged parameter values $\bar{\boldsymbol{\theta}}$ as shown in the bottom part of Fig. 1.4. We will show next that the conditions for this to happen are, in general, impossible to satisfy.

The second-order sufficient conditions of optimality that need to be satisfied *jointly* by the estimation and optimization problems are

$$
\begin{aligned}
&\frac{\partial J^{\mathrm{id}}}{\partial \boldsymbol{\theta}}\left(\mathbf{y}_{\mathrm{p}}\left(\mathbf{u}_{\mathrm{p}}^{*}\right), \mathbf{y}\left(\mathbf{u}_{\mathrm{p}}^{*}, \bar{\boldsymbol{\theta}}\right)\right)=\mathbf{0} \\
&\frac{\partial^{2} J^{\mathrm{id}}}{\partial \boldsymbol{\theta}^{2}}\left(\mathbf{y}_{\mathrm{p}}\left(\mathbf{u}_{\mathrm{p}}^{*}\right), \mathbf{y}\left(\mathbf{u}_{\mathrm{p}}^{*}, \bar{\boldsymbol{\theta}}\right)\right)>\mathbf{0} \\
&G_{i}\left(\mathbf{u}_{\mathrm{p}}^{*}, \bar{\boldsymbol{\theta}}\right)=0 \quad i \in A\left(\mathbf{u}_{\mathrm{p}}^{*}\right) \\
&G_{i}\left(\mathbf{u}_{\mathrm{p}}^{*}, \bar{\boldsymbol{\theta}}\right)<0 \quad i \notin A\left(\mathbf{u}_{\mathrm{p}}^{*}\right) \\
&\nabla_{\mathrm{r}}^{2} \Phi\left(\mathbf{u}_{\mathrm{p}}^{*}, \bar{\boldsymbol{\theta}}\right)>\mathbf{0}
\end{aligned}
\quad [1.13]
$$

where $J_k^{\mathrm{id}} = \|\mathbf{y}_{\mathrm{p}}(\mathbf{u}_k^*) - \mathbf{y}(\mathbf{u}_k^*, \boldsymbol{\theta})\|$ represent the cost function of the identification problem at iteration k (here formulated as the least-squares minimization of the difference between predicted and measured outputs), $A(\mathbf{u}_{\mathrm{p}}^*)$ represents the active set and $\nabla_{\mathrm{r}}^2 \Phi$ the reduced Hessian of the objective function defined as follows: if $\mathbf{Z}$ denotes the null space of the Jacobian matrix of the active constraints and $L = \Phi + \boldsymbol{\nu}^{\mathrm{T}}\mathbf{G}$ the Lagrangian of the optimization problem, then the reduced Hessian is $\nabla_{\mathrm{r}}^2 \Phi = \mathbf{Z}^{\mathrm{T}} \left[\frac{\partial^2 L}{\partial \mathbf{u}^2}\right] \mathbf{Z}$. The first two conditions correspond to the parameter estimation problem, while the other three conditions are linked to the optimization problem. These conditions include both equalities and inequalities, which all depend on the values of $\bar{\boldsymbol{\theta}}$. By itself, the set of equalities in the first condition uses up all the n_θ degrees of freedom, where n_θ denotes the number of model parameters that are estimated. Note that $\mathbf{u}_{\mathrm{p}}^*$ are not degrees of freedom as they correspond to the plant optimum and are therefore fixed. Hence, it is impossible, in general, to satisfy the remaining equality constraints. Furthermore, some of the inequality constraints might also be violated.

Figure 1.5 illustrates through a simulated example that the iterative scheme does not converge to the plant optimum. The two-step approach is applied to optimize a CSTR in which the following three reactions take place (Williams and Otto, 1960):

$$\begin{aligned} A + B &\rightarrow C \\ B + C &\rightarrow P + E \\ C + P &\rightarrow G \end{aligned}$$

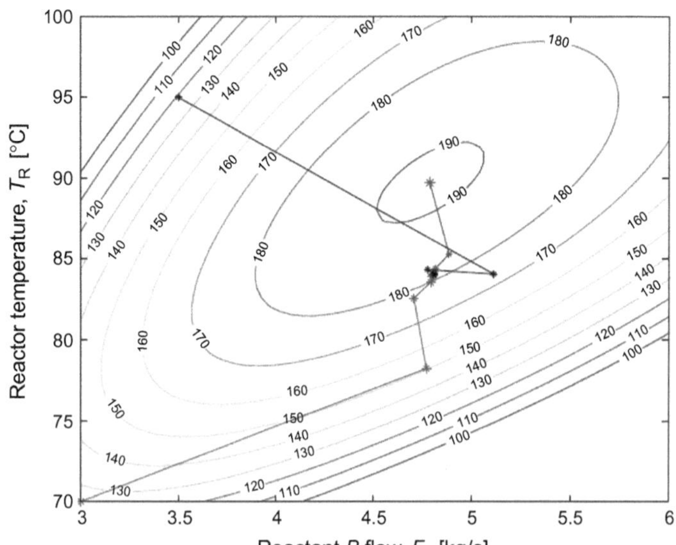

Figure 1.5 Convergence of the two-step RTO scheme to a fixed point that is not the plant optimum (Marchetti, 2009). (For color version of this figure, the reader is referred to the online version of this chapter.)

The model considers only the following two reactions:

$$A + 2B \rightarrow P + E$$
$$A + B + P \rightarrow G$$

but the corresponding kinetic parameters can be adjusted. The inputs are the reactor temperature and the feed rate of one of the reactants. Figure 1.5 shows the contour lines for the plant with the plant optimum in the middle, where the RTO scheme should converge.

With the two-step approach, the kinetic constants of the two modeled reactions are refined iteratively. The updated values are used for the subsequent model-based optimization step, where new values for the steady-state reactor temperature and reactant B flow rate are determined. For three different initial values of the inputs, the scheme converges to the same operating point, which is not the plant optimum. Note that, even when starting at the plant optimum, the algorithm wanders away and converges to a fixed point of the iterative scheme. Hence, the model at hand is not adequate to be used with the two-step approach.

5.4. Modifier-adaptation approach

5.4.1 Basic idea

The modifier-adaptation approach uses measurements in a very different manner than the two-step approach. While for the latter the objective is to match model and process outputs in the hope that the corresponding optimization problems will have matching NCOs, the modifier-adaptation method avoids the parameter identification stage entirely. For this purpose, the optimization problem is modified by the addition of modifier terms to the cost and constraint functions (Marchetti et al., 2009). Intuitively, one sees that, as the NCOs involve (i) the constraints and (ii) the gradients of the cost and constraint functions, the modifiers need to include the deviations between predicted and measured constraints and predicted and measured gradients. With such modifiers, it can be ensured that, upon convergence, the NCOs of the modified problem will match those of the plant. So far, modifier adaptation has been developed for static optimization. It has been proposed to modify the optimization problem as follows:

$$\mathbf{u}_{k+1}^* = \underset{\mathbf{u}}{\text{argmin}} \left\{ \Phi_{\text{m}}(\mathbf{u}) := \Phi(\mathbf{u}) + \left(\left.\frac{\partial \Phi_{\text{p}}}{\partial \mathbf{u}}\right|_{\mathbf{u}_k^*} - \left.\frac{\partial \Phi}{\partial \mathbf{u}}\right|_{\mathbf{u}_k^*} \right) (\mathbf{u} - \mathbf{u}_k^*) \right\}$$

$$\text{s.t.} \quad \mathbf{G}_{\text{m}}(\mathbf{u}) := \mathbf{G}(\mathbf{u}) + \left(\mathbf{G}_{\text{p}}(\mathbf{u}_k^*) - \mathbf{G}(\mathbf{u}_k^*)\right) + \left(\left.\frac{\partial \mathbf{G}_{\text{p}}}{\partial \mathbf{u}}\right|_{\mathbf{u}_k^*} - \left.\frac{\partial \mathbf{G}}{\partial \mathbf{u}}\right|_{\mathbf{u}_k^*} \right) (\mathbf{u} - \mathbf{u}_k^*) \leq \mathbf{0} \qquad [1.14]$$

The optimal inputs computed at iteration k are applied to the plant. The constraints are measured (this is generally the case) and the plant gradient for the cost and the constraints are estimated (which represents a real challenge). The cost and constraint functions are modified by adding zeroth- and first-order correction terms as illustrated for a single constraint in Fig. 1.6. When the optimal inputs $\mathbf{u}_k^*$ are applied to the plant, deviations are observed between the predicted and the measured values of the constraint, that is, $\boldsymbol{\varepsilon}_k = \mathbf{G}_{\text{p}}(\mathbf{u}_k^*) - \mathbf{G}(\mathbf{u}_k^*)$, and also between the predicted and the actual values of the slope, that is, $\boldsymbol{\Lambda}_k^{\text{G}} = \left.\frac{\partial \mathbf{G}_{\text{p}}}{\partial \mathbf{u}}\right|_{\mathbf{u}_k^*} - \left.\frac{\partial \mathbf{G}}{\partial \mathbf{u}}\right|_{\mathbf{u}_k^*}$. These differences are used to both shift the value and adjust the slope of the constraint function. Similar modifications are performed for the cost function, though zeroth-order correction is not necessary, as shifting the value of the cost function does not change the location of its minimizer.

Clearly, the challenge is in estimating the plant gradients. Gradients are necessary for ensuring that, upon convergence, the NCOs of the modified optimization problem match those of the plant. Fortunately, in many cases, constraint shifting by itself achieves most of the optimization potential (Srinivasan et al., 2001); in fact, it is exact when the optimal solution is fully determined by active constraints, that is, when the number of active

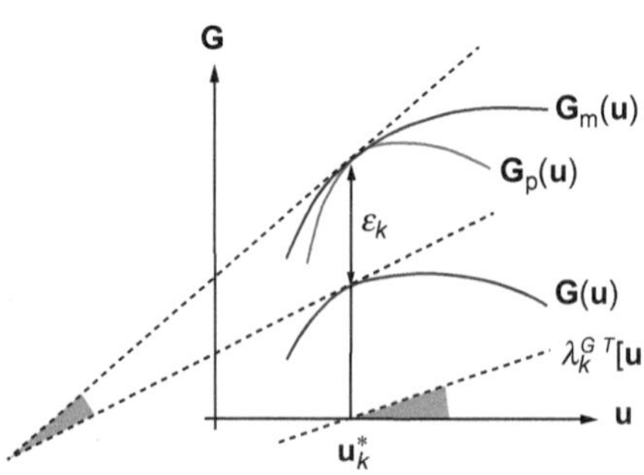

Figure 1.6 Adaptation of the single constraint G at iteration *k*. *Reprinted from Marchetti et al. (2009) with permission of American Chemical Society.* (For color version of this figure, the reader is referred to the online version of this chapter.)

constraints equals the number of inputs. In this case, the implementation is largely simplified, as only the modifier terms $\boldsymbol{\varepsilon}_k = \mathbf{G}_p(\mathbf{u}_k^*) - \mathbf{G}(\mathbf{u}_k^*)$ are required (Marchetti, 2009), and constraint adaptation can be written as

$$\begin{aligned} \mathbf{u}_{k+1}^* &= \underset{\mathbf{u}}{\arg\min} \quad \Phi(\mathbf{u}) \\ \text{s.t.} \quad & \mathbf{G}_m(\mathbf{u}) := \mathbf{G}(\mathbf{u}) + \left(\mathbf{G}_p\left(\mathbf{u}_k^*\right) - \mathbf{G}\left(\mathbf{u}_k^*\right)\right) \leq \mathbf{0} \end{aligned} \quad [1.15]$$

In any case, constraint adaptation is sufficient for enforcing feasibility upon convergence. Figure 1.7 depicts the philosophy of the modifier-adaptation strategy. The adaptation is performed at the level of the optimization problem, which computes the updated inputs.

5.4.2 Model adequacy

We consider the same example and the same two-reaction model as was used previously with the two-step approach, but we now use a RTO scheme that modifies the cost and constraint functions. This example shows that the concept of model adequacy is linked to the optimization approach.

At each iteration, the KKT modifiers are computed from the difference between measured and predicted values of the KKT elements. Note that the KKT modifiers are not computed through optimization. The optimality conditions for this RTO scheme read:

$$\nabla_r^2 \Phi\left(\mathbf{u}_p^*, \bar{\boldsymbol{\theta}}\right) > \mathbf{0} \quad [1.16]$$

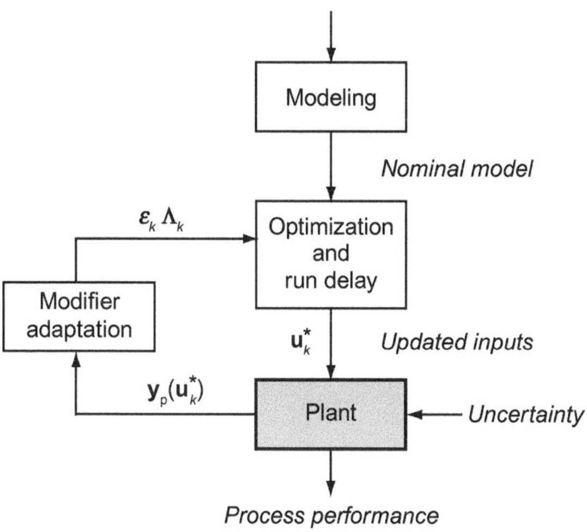

Figure 1.7 Basic idea of modifier adaptation. (For color version of this figure, the reader is referred to the online version of this chapter.)

that is, there are none for the computation of the modifiers, and only a condition on the sign of the reduced Hessian as the first-order NCO are satisfied by construction of the modifiers. Hence, the model is adequate for use with the modifier-adaptation scheme, which is confirmed by the simulation results shown in Fig. 1.8, for which the full modifier-adaptation algorithm of Eq. (1.14) is implemented.

5.5. Self-optimizing approaches

5.5.1 Basic idea

The general idea is to recast the optimization problem as a classical control problem for which the inputs, generally initialized as the model-based optimal inputs, are directly updated through an appropriate control law. In classical control, the distinction between controlled variables (CVs) and manipulated variables (MVs) is quite clear and set points or trajectories to track are part of the problem definition; hence, in classical control, the challenge lies in the choice of the control strategy and the design of the corresponding controller. In self-optimizing control, the real challenge is neither in the choice of control strategy nor in the design of the controller but rather in (i) the definition of the appropriate CVs, (ii) the choice of the

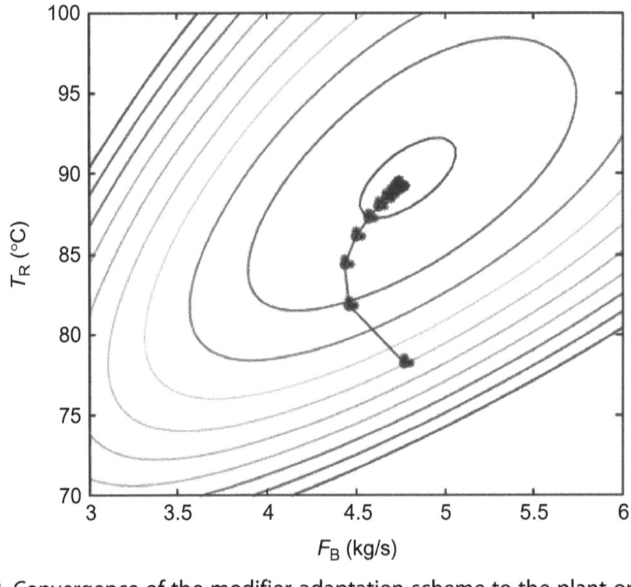

Figure 1.8 Convergence of the modifier-adaptation scheme to the plant optimum for the Williams–Otto reactor (Marchetti, 2009). (For color version of this figure, the reader is referred to the online version of this chapter.)

MVs, (iii) the pairing between MVs and CVs, and (iv) the definition of the set points. The optimization objective would be a natural CV if its set point were known. The various self-optimizing approaches differ in the choice of the CVs, while in general all methods use simple controllers at the implementation level. For instance, with the method labeled "self-optimizing control," one possible choice for the CVs lies in the null space of the sensitivity matrix of the optimal outputs with respect to the uncertain parameters (hence, the source of uncertainty needs to be known) (Alstad and Skogestad, 2007). When there are more outputs than the number of inputs and uncertain parameters together, choosing the CVs as proposed ensures that these CVs are locally insensitive to uncertainty. Hence, these CVs can be controlled at constant set points that correspond to their nominal optimal values by manipulating the inputs of the optimization problem. Figure 1.9 illustrates the information flow of self-optimizing approaches. The effect of uncertainty is rejected by appropriate choice of the control strategy.

5.5.2 NCO tracking

Thereafter, emphasis will be given to NCO tracking (François et al., 2005; Srinivasan and Bonvin, 2007). One consequence of uncertainty is that the optimal inputs computed using the model will not be able to meet the plant NCOs. With NCO tracking, the CVs correspond to measurements or

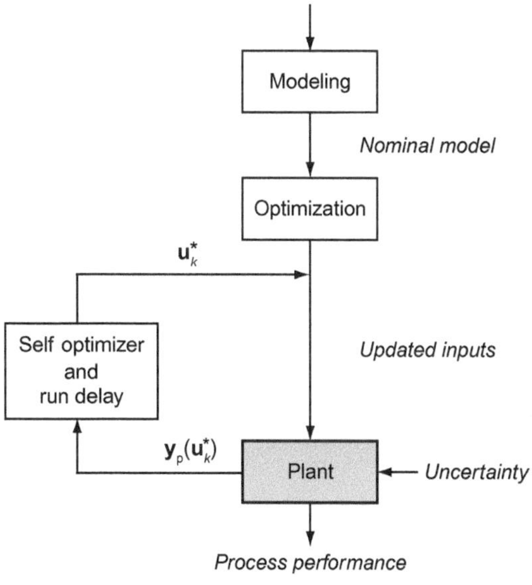

Figure 1.9 Basic idea of self-optimizing approaches. (For color version of this figure, the reader is referred to the online version of this chapter.)

estimates of the plant NCOs, and the set points are the ideal values **0**. Controlling the plant NCOs to zero is indeed an indirect way of solving the optimization problem for the plant, at least in the sense of the first-order NCOs.

Though also applicable to steady-state optimization problems, NCO-tracking exploits its full potential when applied to dynamic optimization problems. In the dynamic case, the NCOs result from application of PMP and encompass four parts: (i) the path constraints, (ii) the path sensitivities, (iii) the terminal constraints, and (iv) the terminal sensitivities. Each degree of freedom of the optimal input profiles satisfies one element in these four parts. Hence, any arc of the optimal solution involves a tracking problem, while time-invariant parameters such as switching times also need to be adapted. To make this problem tractable, NCO tracking introduces the concept of "model of the solution." This concept is key since controlling the NCOs is not a trivial problem. The development of a solution model involves three steps:

1. Characterize the optimal solution in terms of the types and sequence of arcs (typically using the available plant model and numerical optimization).
2. Select a finite set of parameters to represent the input profiles and formulate the NCOs for this choice of degrees of freedom. Pair the MVs and the NCOs to form a multivariable control problem.
3. Perform a robustness analysis to ensure that the nominal optimal solution remains structurally valid in presence of uncertainty, that is, it has the same types and sequence of arcs. If this is not the case, it is necessary to rethink the structure of the solution model and repeat the procedure.

As the solution model formally considers the different parts of the NCOs that need to be enforced for optimality, different control problems will result. A path constraint is often enforced on-line via constraint control, while a path sensitivity is more difficult to control as it requires the knowledge of the adjoint variables. The terminal constraints and sensitivities call for prediction, which is best done using a model, or else, they can be met iteratively over several runs. One of the strength of the approach is that, to ease implementation, it is almost always possible to use simpler profiles for approximating the input profiles, and the approximations introduced at the solution level can be assessed in terms of optimality loss.

6. CASE STUDIES

6.1. Scale-up in specialty chemistry

Short times to market are required in the specialty chemicals industry. One way to reduce this time to market is by skipping the pilot-plant investigations.

Due to scale-related differences in operating conditions, direct extrapolation of conditions obtained in the laboratory is often impossible, especially when terminal objectives must be met and path constraints respected. In fact, ensuring feasibility at the industrial scale is of paramount importance. This section presents an example for which run-to-run control allows meeting production requirements over a few batches.

6.1.1 Problem formulation

Consider the following parallel reaction scheme (Marchetti et al., 2006):

$$A+B\rightarrow C,\quad 2B\rightarrow D. \qquad [1.17]$$

The desired product is C, while D is undesired. The reactions are exothermic. A jacketed reactor of 7.5 m^3 will be used in production, while a 1-L reactor was used in the laboratory. This reaction scheme represents one step of a rather long synthesis route, and the reactor assigned to this step is part of a multi-purpose plant.

The manipulated inputs are the feed rate $F(t)$ and the flow rate of coolant through the jacket $F_j(t)$. The operational requirements are

$$\begin{aligned} &T_j(t)\geq 10°\mathrm{C} \\ &y_D(t_f)=\frac{2n_D(t_f)}{n_C(t_f)+2n_D(t_f)}\leq 0.18 \end{aligned} \qquad [1.18]$$

where n_C and n_D denote the numbers of moles of C and D in the reactor, respectively.

6.1.2 Laboratory recipe

The recipe obtained in the laboratory proposes to initially fill the reactor with A, and then to feed B at some constant feed rate $\bar{F}$, while maintaining the reactor isothermal at $T_r=40\,°\mathrm{C}$. As cooling is not an issue for the laboratory reactor equipped with an efficient jacket, experiments were carried out with a scale-down approach, that is, the cooling rate was artificially limited so as to anticipate the limited cooling capacity of the industrial reactor. Scaling down is performed by the introduction of a constraint that limits the cooling capacity; for this, the maximal cooling capacity of the industrial reactor is simply divided by the scale-up factor:

$$\left[q_{c,\max}\right]_{\mathrm{lab}}=\frac{\left[\left(T_r-T_{j,\min}\right)\mathrm{UA}\right]_{\mathrm{prod}}}{r} \qquad [1.19]$$

Table 1.2 Laboratory recipe for the scale-up problem

Parameters of the recipe		**Experimental results**
$T_r = 40\,°C$	$c_{B_{in}} = 5\,\text{mol/L}$	$n_C(t_f) = 0.346$ mol
$c_{A_0} = 0.5\,\text{mol/L}$	$c_{B_0} = 0\,\text{mol/L}$	$y_D(t_f) = 0.1706$
$V_0 = 1$ L	$t_f = 240$ min	$\max_t q_c(t_f) = 182.6\,\text{J/min}$
$\bar{F} = 4 \times 10^{-4}\,\text{L/min}$		

where $r = 5000$ is the scale-up factor and $UA = 3.7 \times 10^4\,\text{J/(min °C)}$ the estimated heat-transfer capacity of the production reactor. With $T_r - T_{j,\min} = 30°C$, the maximal cooling rate is 222 J/min. Table 1.2 summarizes the key parameters of the laboratory recipe and the corresponding experimental results.

6.1.3 Scale-up seen as a control problem

The recipe is characterized by a set of parameters $\boldsymbol{\rho}$ and the time-varying variables $\mathbf{u}(t)$. For example, the parameter vector $\boldsymbol{\rho}$ could include the feed concentration, the initial conditions and the amount of catalyst, while the profiles $\mathbf{u}(t)$ may correspond to the feed rate and the flow rate of coolant through the jacket.

The first step consists in selecting MVs and CVs. The profiles $\mathbf{u}(t)$ are parameterized as time-varying arcs and switching times between the various arcs. The MVs encompass a certain number of arcs $\boldsymbol{\eta}(t)$ and the parameters $\boldsymbol{\pi}$ that include the parameters $\boldsymbol{\rho}$ and the switching times. The elements of the laboratory recipe that are not chosen as MVs constitute the fixed part of the recipe and are applied as such to the industrial reactor. The CVs include the run-time outputs $\mathbf{y}(t)$ and the run-end outputs $\mathbf{z}$. The objective is to reach the corresponding set points, $\mathbf{y}_{sp}(t)$ and $\mathbf{z}_{sp}$, after as few batches as possible.

The control scheme is proposed in Fig. 1.10, where $\mathbf{y}(t)$ is controlled online with the feedback controller K and run-to-run with the feedforward ILC controller I. Furthermore, $\mathbf{z}$ is controlled on a run-to-run basis using the run-to-run controller R. As direct input adaptation is performed here for rejecting the effect of uncertainty, this example illustrates one possible application of the method described in Section 5.5, with almost all implementation issues discussed in Section 5.2.

6.1.4 Application to the industrial reactor

Temperature control is typically done via a combined feedforward and feedback scheme. The feedback part implements cascade control, for which the

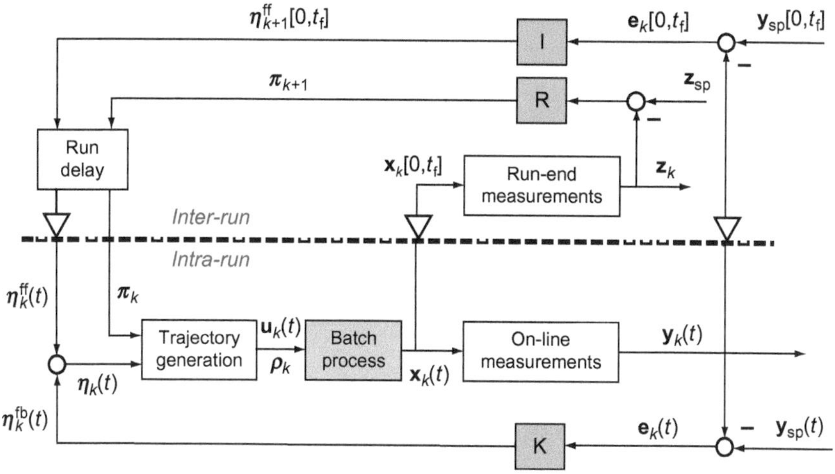

Figure 1.10 Control scheme for scale-up implementation. Notice the distinction between intra-run and inter-run activities. The symbol ∇ represents the concentration/expansion of information between a profile (e.g., $\mathbf{x}_k[0,t_f]$) and an instantaneous value (e.g., $\mathbf{x}_k(t)$). (For color version of this figure, the reader is referred to the online version of this chapter.)

master loop computes the (feedback part of the) jacket temperature set point, $T_{\mathrm{fb,j,sp}}(t)$, while the slave loop adjusts the flow rate of coolant so as to track the jacket temperature set point. The feedforward term for the jacket temperature set point, $T_{\mathrm{ff,j,sp}}(t)$, affects significantly the performance of the temperature control scheme.

The goal of the scale-up is to reproduce in production the final selectivity obtained in the laboratory, while guaranteeing a given productivity of C. For this purpose, the feed rate profile $F[0, t_f]$ is parameterized using the two feed-rate levels F_1 and F_2, each valid over half the batch time, while the final number of moles of C and the final yield represent the run-end CVs. Hence, the control problem can be formulated as follows:

- MV: $\eta(t) = T_{\mathrm{j,sp}}(t)$, $\boldsymbol{\pi} = [F_1 \quad F_2]^T$
- CV: $\mathrm{y}(t) = T_{\mathrm{r}}(t)$, $\mathbf{z} = [\mathrm{n_C}(t_f) \quad \mathrm{y_D}(t_f)]^T$
- SP: $\mathrm{y_{sp}}(t) = 40°\mathrm{C}$, $\mathbf{z}_{\mathrm{sp}} = [1530\ \mathrm{mol} \quad 0.175]^T$

Note that backoffs from the operational constraints are implemented to account for run-time disturbances. The input profiles are updated using (i) the cascade feedback controller K to control the reactor temperature in real time, (ii) the ILC controller I to improve the reactor temperature by adjusting $T_{\mathrm{ff,j,sp}}[0, t_f]$, and (iii) the run-to-run controller R to control $\mathbf{z}$ by adjusting $\boldsymbol{\pi}$. Details regarding the implementation of the different control elements can be found in Marchetti et al. (2006).

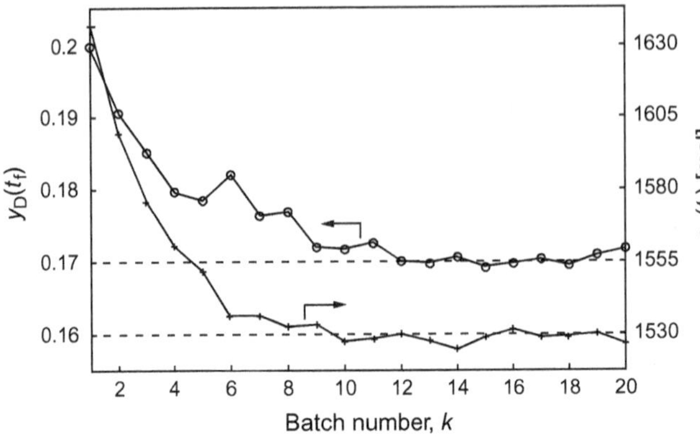

Figure 1.11 Evolution of the yield and the production of C for the large-scale industrial reactor. The two arrows also indicate the time after which adaptation is within the noise level.

6.1.5 Simulation results

The recipe presented below is applied to the 5-m^3 industrial reactor, equipped with a 2.5-m^3 jacket. In addition, uncertainty is introduced in the two kinetic parameters, which are reduced by 25% and 20%, respectively. Also, Gaussian noise with standard deviations of 0.001 mol/L and 0.1 °C is considered for the measurement of the final concentrations of species C and D and for the reactor temperature, respectively. It follows that, for the first run, application of the laboratory recipe with $\boldsymbol{\pi}_1 = [r\bar{F} \;\; r\bar{F}]^T$ results in violation of the final selectivity of D in the first batch. Upon adapting the MVs with the proposed scale-up algorithm, the free parts of the recipe are successfully modified to achieve the production targets for the industrial reactor, as illustrated in Fig. 1.11.

6.2. Solid oxide fuel cell stack

This section describes the application of modifier adaptation to an experimental SOFC stack. Details regarding the model of the stack at hand can be found in Bunin et al. (2012).[1] A SOFC is a system fed with oxygen (air stream) and hydrogen (fuel stream), which react electrochemically to produce electrical power and heat. The fuel cells are assembled in a stack in order to reach the desired voltage. Both the lifetime of cells and the electrical efficiency for a given power demand need to be maximized for SOFC stacks to be more widely used. To control and eventually optimize the stack,

[1] Adapted with permission of Elsevier.

one manipulates the hydrogen and oxygen fluxes and the current that is generated. Furthermore, to assess the stack performance, it is necessary to monitor the power density (which needs to match the power load), the cell potential and fuel utilization (both are bounded to maximize cell lifetime), and the electrical efficiency that represents the optimization objective.

6.2.1 Problem formulation

The constrained model-based optimization problem for maximizing efficiency of the SOFC stack can be written as follows:

$$\begin{aligned} \mathbf{u}^* = \arg\max_{\mathbf{u}} \;& \eta(\mathbf{u},\boldsymbol{\theta}) \\ \text{s.t.}\;& p_{\text{el}}(\mathbf{u},\boldsymbol{\theta}) = p_{\text{el}}^{\text{S}} \\ & U_{\text{cell}}(\mathbf{u},\boldsymbol{\theta}) \geq 0.75\text{V} \\ & \nu(\mathbf{u}) \leq 0.75 \\ & 4 \leq \lambda_{\text{air}}(\mathbf{u}) \leq 7 \\ & u_2 \geq 3.14\text{mL}/(\text{min}\,\text{cm}^2) \\ & u_3 \leq 30\text{A} \end{aligned} \qquad [1.20]$$

where $\mathbf{u} = [u_1 \;\; u_2 \;\; u_3]^T = [\dot{\text{n}}_{\text{O}_2} \;\; \dot{\text{n}}_{\text{H}_2} \;\; \text{I}]^T$ is the vector of manipulated inputs (the molar fluxes of oxygen and hydrogen and the current), $\boldsymbol{\theta}$ the vector of seven uncertain model parameters, $\eta(\mathbf{u},\boldsymbol{\theta})$ the electrical efficiency, $p_{\text{el}}(\mathbf{u},\boldsymbol{\theta})$ the produced power density, p_{el}^{S} the power load, $U_{\text{cell}}(\mathbf{u},\boldsymbol{\theta})$ the cell potential, $\nu(\mathbf{u}) = \frac{N_{\text{cells}} u_3}{2u_2 F}$ the fuel utilization, N_{cells} the number of cells, F Faraday constant, and $\lambda_{\text{air}}(\mathbf{u}) = 2\frac{u_1}{u_2}$ the oxygen-to-hydrogen ratio. Several remarks are in order:

- $\nu(\mathbf{u})$ and $\lambda_{\text{air}}(\mathbf{u})$ are not affected by uncertainty because they are computed from inputs that are known with certainty.
- $p_{\text{el}}(\mathbf{u},\theta)$, $U_{\text{cell}}(\mathbf{u},\theta)$ and $\eta(\mathbf{u},\theta)$ are computed from the model and thus affected by uncertainty.
- The optimization is formulated as a steady-state optimization problem though the system is dynamic. There are two main time scales: (i) the electrochemical time scale, which is almost instantaneous, and (ii) the thermal scale (i.e., the dynamics associated with thermal equilibrium, the SOFC being installed in a furnace) with a settling time of about 30 min.
- The first constraint indicates that the stack has to produce the power required by the user p_{el}^{S}. This value can vary and is measured on-line, but it is not known in advance nor can it be predicted. Hence, the challenge is to track this equality constraint, while maximizing electrical efficiency.
- The lower bound on cell potential prevents the SOFC from accelerated degradation.

- The upper bound on fuel utilization prevents damages to the stack caused by local fuel starvation and re-oxidation of the anode.

6.2.2 RTO via constraint adaptation

Numerical simulation has shown that the optimal solution is determined by active constraints. In fact, the constraint on fuel utilization becomes active at low power loads, while the constraint on cell potential becomes limiting at high power demands. Hence, constraint control is sought for both optimality and safety reasons. Said differently, the solution will always be on the constraint of either fuel utilization or cell potential, but (i) it is impossible to know in advance which constraint should be tracked (as the power load is not known in advance), and (ii) given the value of the power load, the model alone may not be sufficient for choosing the constraint to track.

At the kth iteration, the following optimization problem is solved for $\mathbf{u}_{k+1}^*$ using the modifiers $\varepsilon_k^{p_{\mathrm{el}}}$ and $\varepsilon_k^{U_{\mathrm{cell}}}$ from the previous iteration:

$$\begin{aligned} \mathbf{u}_{k+1}^* = &\arg\max_{\mathbf{u}} \ \eta(\mathbf{u},\boldsymbol{\theta}) \\ \text{s.t.}\ & p_{\mathrm{el}}(\mathbf{u},\boldsymbol{\theta}) + \varepsilon_k^{\mathrm{P_{el}}} = p_{\mathrm{el}}^{\mathrm{S}} \\ & U_{\mathrm{cell}}(\mathbf{u},\boldsymbol{\theta}) + \varepsilon_k^{\mathrm{U_{cell}}} \geq 0.75\mathrm{V} \\ & \nu(\mathbf{u}) \leq 0.75 \\ & 4 \leq \lambda_{\mathrm{air}}(\mathbf{u}) \leq 7 \\ & u_2 \geq 3.14\mathrm{mL}/(\mathrm{min\,cm^2}) \\ & u_3 \leq 30\mathrm{A} \end{aligned} \qquad [1.21]$$

The modifiers are filtered with an exponential filter of gain K. Upon convergence, the solution of the modified optimization problem is guaranteed to satisfy the constraints for the real stack. The modifiers then indicate the errors between experimental and predicted values. The general algorithm proceeds as follows:

i. Set $k=0$ and initialize the modifiers to zero.

ii. Solve the modified optimization problem to obtain the new input values $\mathbf{u}_{k+1}^*$.

iii. Assume convergence if $\|\mathbf{u}_{k+1}^* - \mathbf{u}_k^*\| \leq \delta$, where δ is a user-specified threshold.

iv. Apply these input values and let the system converge to a new steady state.

v. Update the modifiers according to and return to Step (ii).

$$\begin{aligned} \boldsymbol{\varepsilon}_k^{p_{\mathrm{el}}} &= \left(1 - K_{p_{\mathrm{el}}}\right)\boldsymbol{\varepsilon}_{k-1}^{p_{\mathrm{el}}} + K_{p_{\mathrm{el}}}\left[p_{\mathrm{el,p}}\left(\mathbf{u}_k^*\right) - p_{\mathrm{P_{el}}}\left(\mathbf{u}_k^*,\boldsymbol{\theta}\right)\right] \\ \boldsymbol{\varepsilon}_k^{U_{\mathrm{cell}}} &= \left(1 - K_{U_{\mathrm{cell}}}\right)\boldsymbol{\varepsilon}_{k-1}^{U_{\mathrm{cell}}} + K_{U_{\mathrm{cell}}}\left[U_{\mathrm{cell,p}}\left(\mathbf{u}_k^*\right) - U_{\mathrm{cell}}\left(\mathbf{u}_k^*,\boldsymbol{\theta}\right)\right] \end{aligned} \qquad [1.22]$$

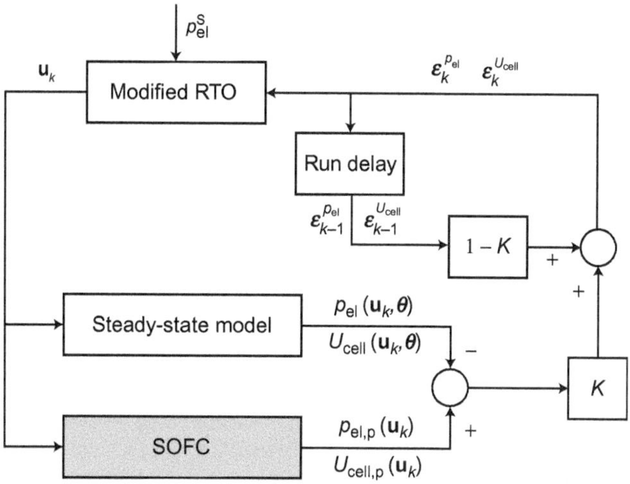

Figure 1.12 Constraint-adaptation scheme for the SOFC stack. (For color version of this figure, the reader is referred to the online version of this chapter.)

As illustrated in Fig. 1.12, the differences between predicted and measured constraints on the power load and on the cell potential are used to modify the RTO problem. Although the system is dynamic, a steady-state model is used, which is justified by the goal of maximizing steady-state performance.

6.2.3 Experimental scenarios

In order to test the ability of the method to enforce maximal electrical efficiency and satisfaction of the constraints despite variable power demand, two different scenarios will be tested, namely, (i) the power demand changes slowly as the system is allowed to reach steady state between two successive changes, and (ii) the power demand changes very fast.

– For scenario (i), the power demand varies as follows:

$$p_{el}^S(t) = \begin{cases} 0.3\dfrac{\mathrm{W}}{\mathrm{cm}^2} & t < 90\,\mathrm{min} \\ 0.38\dfrac{\mathrm{W}}{\mathrm{cm}^2} & 90\,\mathrm{min} \leq t < 180\,\mathrm{min} \\ 0.3\dfrac{\mathrm{W}}{\mathrm{cm}^2} & t \geq 180\,\mathrm{min} \end{cases} \qquad [1.23]$$

Again, note that this information is not known at the implementation level. Constraint adaptation is performed from one steady state to the next using only steady-state measurements.

- For scenario (ii), the power load is changed randomly every 5 min in the same range as for scenario (i). Hence, the system does not have time to reach steady state. RTO is performed every 10 s using on-line measurements. Because the RTO update is much faster than the thermal settling time, the error made by predicting the temperature using a static model will be small and, furthermore, it will be rejected like any other source of uncertainty.

6.2.4 Experimental results

Figures 1.13 and 1.14 illustrate the application of RTO via modifier adaptation to the experimental SOFC stack for slow and fast variations of the power demand, respectively.

The upper left plot of Fig. 1.13 shows that, upon convergence, the RTO scheme meets the active constraint on power demand. The plots of fuel utilization and cell potential indicate that, at low loads, the constraint on fuel utilization gets activated, while at high loads, the constraint on cell potential is reached after a couple of RTO iterations. Finally, the right bottom plot shows that electrical efficiency increases over RTO iterations.

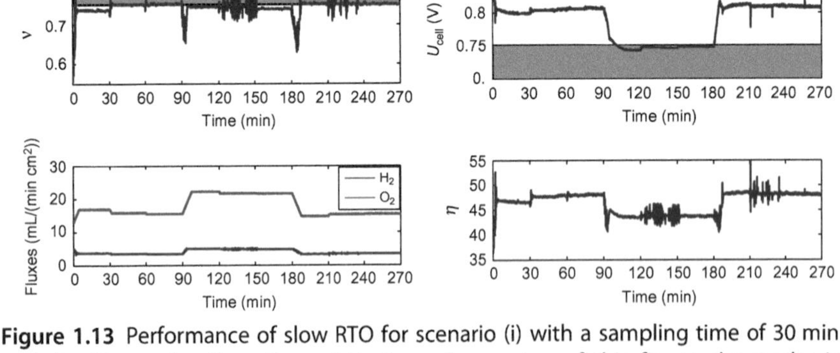

Figure 1.13 Performance of slow RTO for scenario (i) with a sampling time of 30 min and the filter gains $K_{p_{el}} = K_{U_{cell}} = 0.7$. (For color version of this figure, the reader is referred to the online version of this chapter.)

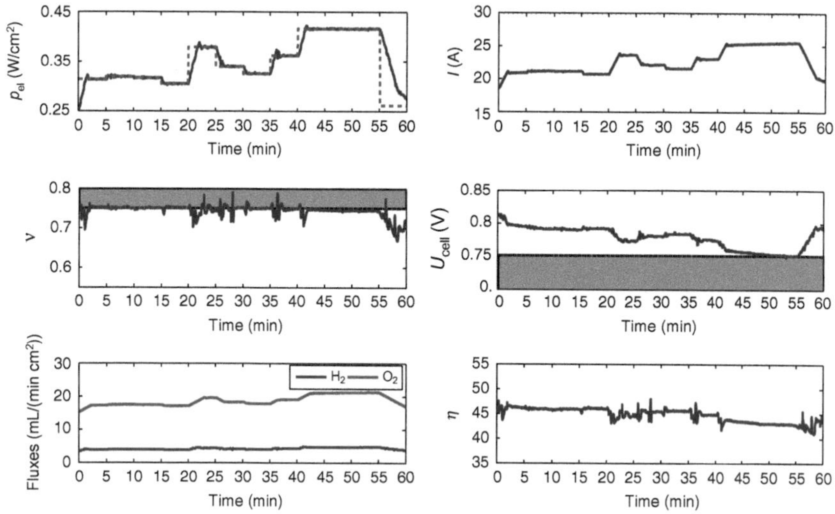

Figure 1.14 Performance of fast RTO for scenario (ii) with a sampling time of 10 s and the filter gains $K_{p_{el}} = 0.85$ and $K_{U_{cell}} = 1.0$. (For color version of this figure, the reader is referred to the online version of this chapter.)

Figure 1.14 illustrates that, with fast RTO, the power load is tracked with much more reactivity. Meanwhile, the constraints on cell potential and fuel utilization are reached quickly, despite the use of inaccurate temperature predictions.

This case study illustrates the use of the strategy discussed in Section 5.4, with the implementation issues of Sections 5.2.2 and 5.2.4.

6.3. Grade transition for polyethylene reactors

This case study considers a fluidized-bed gas-phase polymerization reactor, with several grades of polyethylene being produced in the same equipment by changing the operating conditions. The problem of grade transition is viewed here as a dynamic optimization problem, with the aim of minimizing the transition time or the amount of off-spec products. Model-based optimization is clearly insufficient in this example due to the presence of uncertainty in the form of plant-model mismatch and process disturbances. NCO tracking is used to adapt the arcs and switching times that have been determined through analysis of the nominal solution and construction of a solution model.

6.3.1 Process description

Polymerization of ethylene in a fluidized-bed reactor with a heterogeneous Ziegler–Natta catalyst is considered. Ethylene, hydrogen, inert (nitrogen) and catalyst are fed continuously to the reactor. Recycle gases are pumped through a heat exchanger and back to the bottom of the reactor. As the single pass conversion of ethylene in the reactor is usually low (1–4%), the recycle stream is much larger than the inflow of fresh feed. Excessive pressure and impurities are removed from the system in a bleed stream at the top of the reactor. Fluidized polymer product is removed from the base of the reactor through a discharge valve. The removal rate of product is adjusted by a bed-level controller that keeps the polymer mass in the reactor at the desired set point. For model-based investigations, a simplified first-principles model is used that is based on the work of McAuley and MacGregor (1991), McAuley et al. (1995), and detailed in Gisnas et al. (2004). Figure 1.15 depicts the fluidized-bed reactor considered in this section.

6.3.2 The grade transition problem

During steady-state production of polyethylene, the operating conditions are chosen to maximize the outflow rate of polymer of desired grade, while meeting operational and safety requirements.

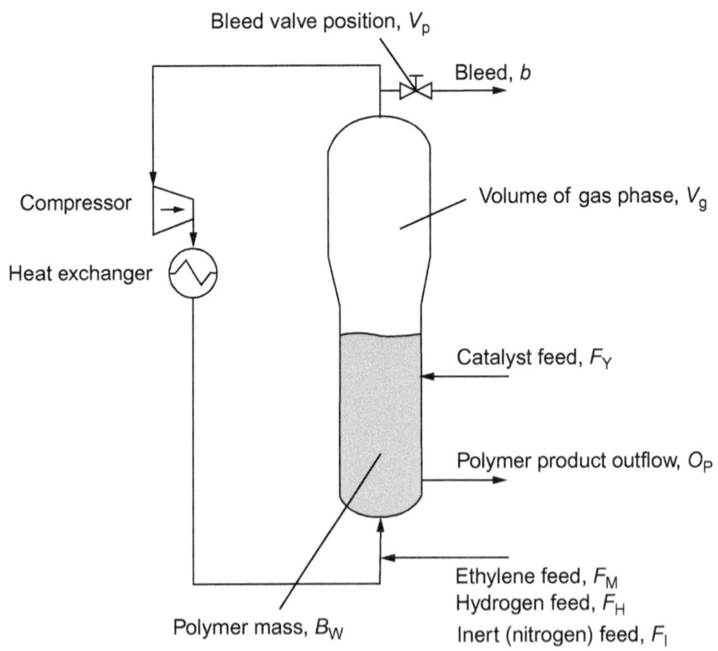

Figure 1.15 Gas-phase fluidized-bed polyethylene reactor.

Table 1.3 Optimal operating conditions and active constraints for grades *A* and *B*, as well as upper and lower bounds used in steady-state optimization

	A	*B*	Lower bound	Upper bound	Set to meet
$MI_{c,ref}$ (g/10 min)	0.009	0.09			
$B_{w,ref}$ (10^3 kg)	70	70			
P (atm)	20	20			
F_H (kg/h)	1.1	15	0	70	$MI_{c,ref}$
F_I (kg/h)	495	281	0	500	P_{ref}
F_M (10^3 kg/h)	30	30	0	**30**	$F_{M,max}$
F_Y (10^{-3} kmol/h)	10	10	0	**10**	$F_{Y,max}$
V_p	0.5	0.5	**0.5**	1	$V_{p,min}$
O_p (10^3 kg/h)	29.86	29.84	21	39	$B_{w,ref}$

6.3.2.1 Analysis of the sets of optimal conditions for grades *A* and *B*

The optimal operating conditions for the two grades *A* and *B* have been determined by solving a static optimization problem (Gisnas et al., 2004). These conditions are presented in Table 1.3 along with the upper and lower bounds used in the optimization.

V_p is maintained at $V_{p,min}=0.5$ to have a nonzero bleed at steady state to be able to handle impurities. Clearly, F_M and F_Y are set to their maximal values, as this maximizes the production of polyethylene and productivity, respectively. F_I is set to have the pressure at its lower bound of 20 atm to minimize the waste of monomer through the bleed. Finally, F_H is determined from the melt index requirement, and O_P is set to keep the polymer mass at its reference value. Hence, for steady-state optimal operation, the six input variables are determined by six active constraints or references.

6.3.2.2 Grade transition as a dynamic optimization problem

The objective is to minimize the transition time t_{trans} to go from grade *A* (with low melt index) to grade *B* (with high melt index). Among the six inputs, only F_H and O_P are considered as decision variables, while the other four are kept at active bounds (see quantities in bold in Table 1.3; note that F_I is fixed at its lower bound to keep the pressure as low as possible during transition). Note also that the polymer mass B_w is allowed to vary. The dynamic optimization problem is stated mathematically as (Bonvin et al., 2005)[2]:

[2] Adapted with permission of Elsevier.

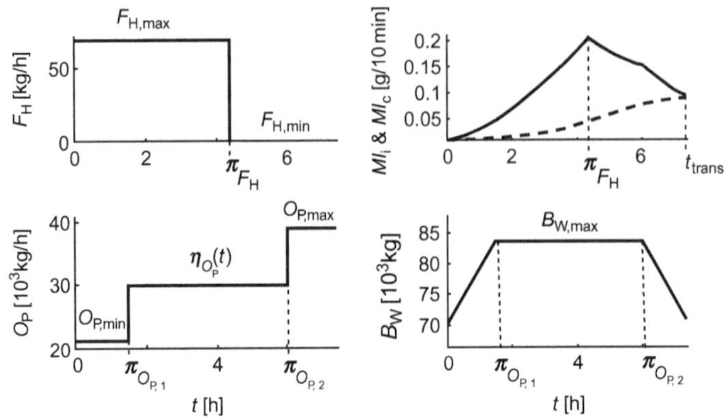

Figure 1.16 Optimal profiles for the transition $A \to B$ (MI_i solid line, MI_c dashed line).

$$
\begin{array}{ll}
\min\limits_{F_H(t), O_P(t), t_{trans}} & J = t_{trans} \\
\text{s.t.} & \text{dynamic equations} \\
& F_{H,\min} \le F_H(t) \le F_{H,\max} \\
& O_{P,\min} \le O_P(t) \le O_{P,\max} \\
& B_{w,\min} \le B_w(t) \le B_{w,\max} \\
& MI_c(t_{trans}) = MI_{c,ref} \\
& MI_i(t_{trans}) = MI_{c,ref} \\
& B_w(t_{trans}) = B_{w,ref}
\end{array}
\qquad [1.24]
$$

where MI_c and MI_i are the cumulated and instantaneous melt indexes, respectively.

6.3.3 The model of the solution

The nominal solution of the dynamic optimization problem is depicted in Fig. 1.16. This solution can be interpreted intuitively as follows:

- F_H is maximal initially in order to increase MI_i as quickly as possible through an increase of $[H_2]$. F_H then switches to its lower bound to meet the terminal constraint on MI_i.
- O_P is minimal initially to help increase MI_i, which can be accomplished through a decrease of [M]. For this, more catalyst is needed, that is, Y is increased. This is achieved by removing less catalyst with the product, which explains why the outlet valve is closed, $O_P = O_{P,min}$. When the outlet valve is closed, the polymer mass increases until B_W reaches its

upper bound. Then, O_P is adjusted to keep this constraint active, which gives the second arc $\eta_{O_P}(t)$. Finally, O_P is maximal in order to decrease the polymer mass and meet the corresponding terminal constraint on B_w.

This analysis of the nominal solution underlines the intrinsic links between the MVs and the path and terminal constraints of the dynamic optimization problem. Applying directly the profiles depicted in Fig. 1.16 will not be optimal, because of plant-model mismatch and disturbances. However, once it has been verified in simulation that uncertainty does not modify the structure of the optimal solution, that is, the types and the sequence of arcs, this information can be used to design the NCO-tracking scheme, which will adapt the profiles to make them match the plant NCOs.

To generate the solution model, the nominal optimal solution is analyzed arc by arc and the inputs are parameterized accordingly; then, the MVs and CVs are selected and an appropriate paring is proposed. The procedure is as follows:

1. Input parameterization
   a. The nominal solution presented in Fig. 1.16 consists of constraint-seeking arcs that are determined by either input bounds or the state constraint B_w, but it does not contain sensitivity-seeking arcs.
   b. The adjustable free parts of the input profiles are the state-constrained arc $\eta_{O_P}(t)$ and the switching times.
   c. As there are no sensitivity-seeking arcs, the parameter vector $\boldsymbol{\pi}$ contains only the switching times π_{F_H}, $\pi_{O_{P,1}}$ and $\pi_{O_{P,2}}$ and the final time t_{trans}.
2. Pairing MVs and CVs
   a. The MV $\eta_{O_P}(t)$ is linked to the state constraint $B_w(t) = B_{w,max}$. The parameter $\pi_{O_{P,1}}$ is determined implicitly upon $B_w(t)$ reaching $B_{w,max}$.
   b. The remaining parameters π_{F_H}, $\pi_{O_{P,2}}$ and t_{trans} are linked to the terminal constraints on $MI_i(t_{trans})$, $B_w(t_{trans})$ and $MI_c(t_{trans})$, respectively.

6.3.4 NCO-tracking scheme

Using the pairing of MVs and CVs, it is straightforward to design a control scheme that enforces the plant NCOs. The following on-line control laws are proposed:

$$F_H(t) = \begin{cases} F_{H,max} & \text{for } 0 \le t < \pi_{F_H} \\ F_{H,min} & \text{for } \pi_{F_H} \le t < t_{trans} \end{cases}$$
$$O_P(t) = \begin{cases} O_{P,min} & \text{for } 0 \le t < \pi_{O_{P,1}} \\ K_{\eta_{O_P}}(B_{w,max} - B_w(t) & \text{for } \pi_{O_{P,1}} \le t < \pi_{O_{P,2}} \\ O_{P,max} & \text{for } \pi_{O_{P,2}} \le t < t_{trans} \end{cases} \quad [1.25]$$

$\pi_{O_{P,1}}$ is determined implicitly upon $B_w(t)$ reaching $B_{w,max}$, while the remaining time-invariant parameters can be adapted using the following run-to-run control laws:

$$\begin{aligned}\pi_{F_H} &= \mathfrak{R}_{\pi_{F_H}}\left(MI_{c,ref} - MI_i(t_{trans})\right)\\ \pi_{O_{P,2}} &= \mathfrak{R}_{\pi_{O_{P,2}}}\left(B_{w,ref} - B_w(t_{trans})\right)\\ t_{trans} &= \mathfrak{R}_{t_{trans}}\left(MI_{c,ref} - MI_i(t_{trans})\right)\end{aligned} \quad [1.26]$$

Combined on-line and off-line control will adapt the profile, over a few batches, to match the plant NCOs. Figure 1.17 depicts the NCO-tracking scheme.

6.3.5 Simulation results

Uncertainty is present in the form of time-varying kinetic parameters, which might correspond to a variation of catalyst efficiency with time. This information is only used to compute the "ideal" minimal transition time, $J^* = 7.36$ h. Table 1.4 summarizes the results. As some of the constraints are violated during the first two runs, for the purpose of comparison, the cost values given in Table 1.4 are artificially penalized for constraint violations (see Bonvin et al., 2005). Convergence to the optimal solution is

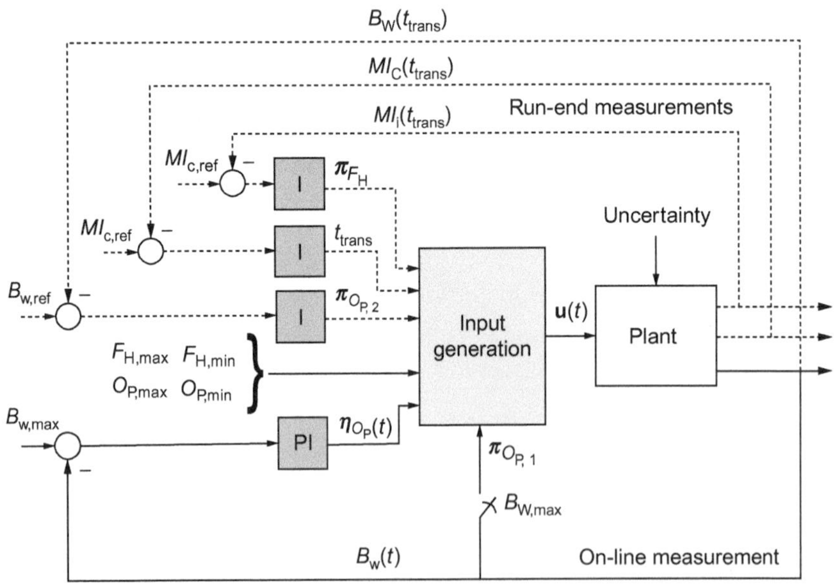

Figure 1.17 NCO-tracking scheme for the grade transition problem. The solid and dashed lines correspond to on-line and run-to-run control, respectively.

Table 1.4 Adaptation results for the grade transition problem

Run number	$\frac{MI_c(t_{trans})}{MI_{c,ref}}$	$\frac{MI_i(t_{trans})}{MI_{c,ref}}$	$\frac{B_w(t_{trans})}{B_{w,ref}}$	t_{trans}[h]	J[h]
1	1.078	1.089	0.999	7.45	10.39
2	1.033	1.045	1.008	7.39	8.88
3	1	1	1	7.36	7.36
10	1	1	1	7.36	7.36

achieved within three runs. Note that considerable cost improvement is achieved after two runs already.

This case study has shown the value of MBO techniques for grade transition problems. A combination of run-to-run and on-line control has been used. Run-to-run control is possible as grade transitions are usually repeated. However, in the presence of multiple grades, it can happen that a given transition is only repeated infrequently. Hence, it is of great interest to be able to meet the terminal constraints, which are most important from a cost point of view, on-line as proposed in Srinivasan and Bonvin (2004). With regard to the MBO techniques discussed in Section 5, the proposed NCO scheme belongs to Section 5.5 and it uses decentralized control.

6.4. Industrial batch polymerization process

The fourth case study illustrates the use of NCO tracking for the optimization of an industrial reactor for the copolymerization of acrylamide (François et al., 2004).[3] As the polymer is repeatedly produced in a batch reactor, run-to-run NCO tracking (using run-end measurements) is applied.

6.4.1 A brief description of the process

The 1-ton industrial reactor investigated in this section is dedicated to the inverse-emulsion copolymerization of acrylamide and quaternary ammonium cationic monomers, a heterogeneous water-in-oil polymerization process.

Nucleation and polymerization are confined to the aqueous monomer droplets, while the polymerization follows a free-radical mechanism. Table 1.5 summarizes the reactions that are known to occur.

A tendency model capable of predicting the conversion and the average molecular weight has been developed. The model parameters have been fitted to match observed data. For reasons of confidentiality, this tendency model

[3] Reprinted and adapted with permission of American Chemical Society.

Table 1.5 Main reactions in the inverse-emulsion process

Oil-phase reactions • initiation by initiator decomposition • reactions of primary radicals • propagation reactions
Transfer between phases • initiator • comonomers • primary radicals
Aqueous-phase reactions • reactions of primary radicals • propagation reactions • unimacromolecular termination with emulsifier • reactions of emulsifier radicals • transfer to monomer • addition to terminal double bond • termination by disproportionation

cannot be presented here. Although this model represents a valuable tool for performing model-based investigations, it is not sufficiently accurate to be used on its own. In addition to structural plant-model mismatch, certain disturbances are nearly impossible to avoid or predict. For instance, the efficiency of the initiator and the efficiency of initiation by emulsifier radicals can vary significantly between batches because of the residual oxygen concentration at the outset of the reaction. Chain transfer agents and reticulants are also added to help control the molecular weight distribution. These small variations in recipe are not incorporated in the tendency model. Hence, optimization of this process clearly calls for the use of measurement-based techniques.

6.4.2 Nominal optimization of the tendency model

The objective is to minimize the reaction time, while meeting four constraints, namely, (i) the terminal molecular weight $\bar{M}_w(t_f)$ is bounded from below to ensure in-spec production, (ii) the terminal conversion $X(t_f)$ has to exceed a target value X_{min} to ensure total conversion of acrylamide, (iii) heat removal is limited, which is incorporated in the optimization problem by the lower bound $T_{j,in,min}$ on the jacket inlet temperature $T_{j,in}(t)$, and (iv) the reactor temperature $T(t)$ is upper bounded. The MVs are the reactor temperature $T(t)$ and the reaction time t_f. The dynamic optimization problem can be formulated as follows:

$$\begin{array}{ll} \min\limits_{T(t),t_f} & t_f \\ \text{s.t.} & \text{dynamic model} \\ & X(t_f) \geq X_{\min} \\ & \bar{M}_w(t_f) \geq \bar{M}_{w,\min} \\ & T_{j,in}(t) \geq T_{j,in,\min} \\ & T(t) \leq T_{\max} \end{array} \qquad [1.27]$$

This formulation considers determining the *reactor temperature* that minimizes the reaction time. Since an optimal strategy computed this way might require excessive cooling, a lower bound on the jacket inlet temperature is added to the problem.

6.4.3 The model of the solution

The results of nominal optimization are shown in Fig. 1.18, with normalized values of the reactor temperature $T(t)$ and the time t.

The nominal optimal solution consists of two arcs with the following interpretation:

- *Heat removal limitation.* Up to a certain level of conversion, the temperature is limited by heat removal. Initially, the operation is isothermal and corresponds closely to what is used in industrial practice. Also, this first isothermal arc ensures that the terminal constraint on molecular weight will be satisfied as it is mostly determined by the concentration of chain transfer agent.

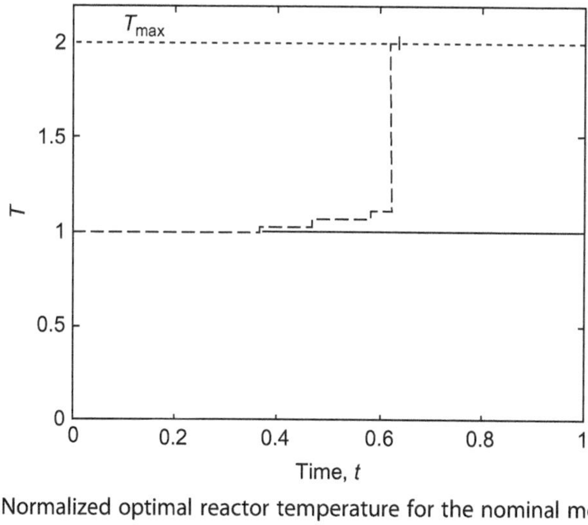

Figure 1.18 Normalized optimal reactor temperature for the nominal model.

- *Intrinsic compromise.* The second arc represents a compromise between reaction speed and quality. The decrease in reaction rate due to smaller monomer concentrations is compensated by an increase in temperature, which accelerates the reaction but decreases molecular weight.

This interpretation of the nominal solution is the basis for the solution model. As operators are reluctant to change the temperature policy during the first part of the batch and the reaction is highly exothermic, it has been decided to:

- Implement the first arc isothermally, with the temperature kept at the value used in industrial practice.
- Implement the second arc adiabatically, that is, without jacket cooling. The reaction mixture is heated up by the reaction, which allows linking the maximal reachable temperature to the amount of reactants (and thus the conversion) at the time of switching.

With this so-called "semi-adiabatic" temperature profile, there are only two degrees of freedom, the switching time between the two arcs, t_{sw} and the final time t_f. The dynamic optimization problem can be rewritten as the following *static* problem:

$$\begin{aligned} \min_{t_f, t_{sw}} \quad & J = t_f \\ & X(t_f) \geq X_{min} \\ & \bar{M}_w(t_f) \geq \bar{M}_{w,min} \\ & T(t_f) \leq T_{max} \end{aligned} \qquad [1.28]$$

This reformulation calls for some remarks:

a. The switching time t_{sw} and the final time t_f are fixed at the beginning of the batch, while performance and constraints are evaluated at batch end. This way, the dynamics are lumped into the static map $(t_{sw}, t_f) \rightarrow \{J, X(t_f), \bar{M}_w(t_f), T(t_f)\}$.

b. Maintaining the temperature constant initially at its current practice value ensures that the heat removal limitation is satisfied. This constraint can thus be removed from the problem formulation.

c. The semi-adiabatic profile ensures that the maximal temperature is reached at batch end.

Because (i) the constraint on the molecular weight is less restrictive than that on the reactor temperature, (ii) the final time is defined upon meeting the desired conversion, and (iii) the terminal constraint on reactor temperature is active at the optimum, the NCOs reduce to the following two conditions:

$$\begin{cases} T(t_f) - T_{\max} = 0 \\ \dfrac{\partial t_f}{\partial t_{sw}} + \nu \dfrac{\partial [T(t_f) - T_{\max}]}{\partial t_{sw}} = 0 \end{cases} \qquad [1.29]$$

where ν is the Lagrange multiplier associated with the constraint on final temperature. The first equation determines the switching time, while the second can be used for computing ν, which, however, is of little interest here.

6.4.4 Industrial results

The solution to the original dynamic optimization problem can be approximated by adjusting the switching time so as to meet the terminal constraint on reactor temperature. This can be implemented using a simple run-to-run controller of gain K, as shown in Fig. 1.19.

Figure 1.20 depicts the application of the method to the optimization of the 1-ton industrial reactor. The first batch is performed using a conservative value of the switching time. The reaction time is significantly reduced after only two batches, without any off-spec product as illustrated in Fig. 1.21 that shows the normalized product viscosity (which correlates well with molecular weight).

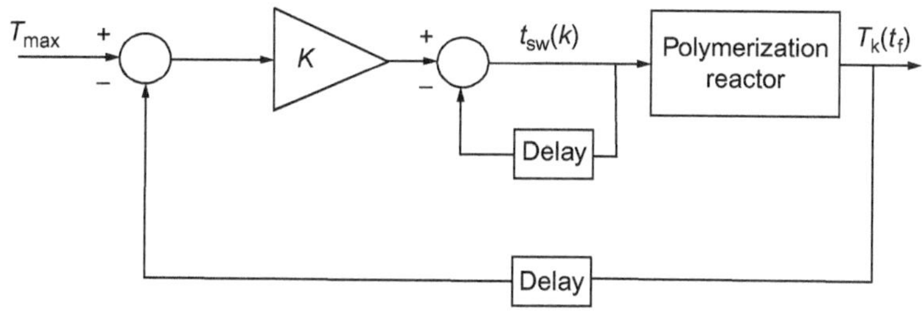

Figure 1.19 Run-to-run NCO-tracking scheme.

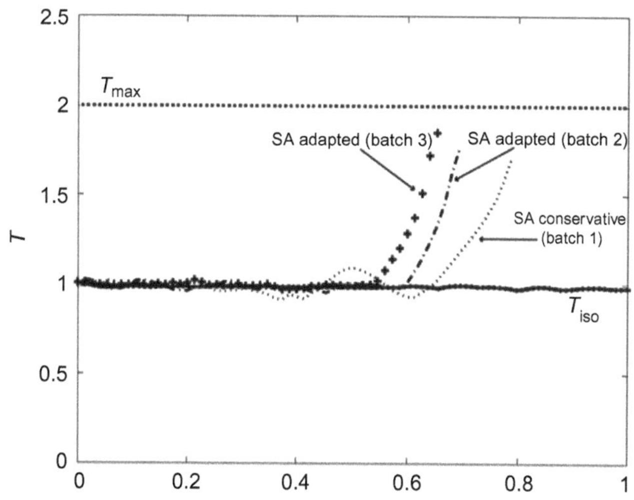

Figure 1.20 Measured temperature profiles for four batches in the 1-ton reactor. Note the significant reduction in reaction time.

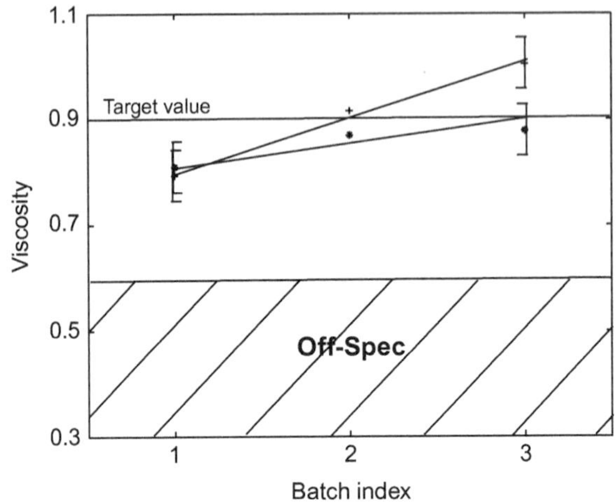

Figure 1.21 Normalized viscosity for the first three batches.

Table 1.6 Run-to-run optimization results for a 1-ton copolymerization reactor

Batch	Strategy	t_{sw}	$T(t_f)$	t_f
–	Isothermal	–	1.00	1.00
1	Semi-adiabatic	0.65	1.70	0.78
2	Semi-adiabatic	0.58	1.78	0.72
3	Semi-adiabatic	0.53	1.85	0.65

Table 1.6 summarizes the adaptation results, highlighting the 35% reduction in reaction time compared to the isothermal policy used in industrial practice. Results could have been even more impressive, but a backoff from the constraint on the final temperature was added and $T_{\max}=1.85$ was used instead of the real constraint value $T_{\max}=2$.

This semi-adiabatic policy has become standard practice for our industrial partner. The same policy has also been implemented, together with the adaptation scheme, to other polymer grades and to larger reactors.

7. CONCLUSIONS

This chapter has shown that incorporating measurements in the optimization framework can help improve the performances of chemical processes when faced with models of limited accuracy. The various MBO methods differ in the way measurements are used and inputs are adjusted to reject the effect of uncertainty. Measurements can be utilized to iteratively (i) update the parameters of the model that is used for optimization, (ii) modify the objective and constraint functions of the optimization problem, and (iii) directly adjust inputs to enforce the NCOs. It has been argued that the two latter techniques have the ability of rejecting the effect of uncertainty in the form of plant-model mismatch and process disturbances.

The use of these MBO methods has been motivated by four common applications: a scale-up problem in specialty chemistry, the steady-state optimization of a fuel cell stack, grade transition in polyethylene reactors, and the dynamic optimization of a batch polymerization reactor. The four case studies include two simulated industrial problems, one experimental setup and one industrial process; they have been optimized using either modifier adaptation or NCO tracking, which highlights the potential of MBO techniques for solving real-life industrial problems.

ACKNOWLEDGMENT

The authors would like to thank the former and present group members at EPFL's Laboratoire d'Automatique who contributed many of the insights and results presented here.

REFERENCES

Alstad V, Skogestad S: Null space method for selecting optimal measurement combinations as controlled variables, *Ind Eng Chem Res* 46(3):846–853, 2007.

Ariyur K, Krstic M: *Real-time optimization by extremum-seeking control,* New York, 2003, John Wiley.

Bazarra MS, Sherali HD, Shetty CM: *Nonlinear programming: theory and algorithms,* ed 2, New York, 1993, John Wiley & Sons.

Biegler LT, Grossmann IE, Westerberg AW: A note on approximation techniques used for process optimization, *Comp Chem Eng* 9:201–206, 1985.

Bonvin D, Srinivasan B, Ruppen D: Dynamic optimization in the batch chemical industry, In *Chemical Process Control-VI,* Tucson, AZ, 2001.

Bonvin D, Bodizs L, Srinivasan B: Optimal grade transition for polyethylene reactors via NCO tracking, *Trans IChemE Part A Chem Eng Res Design* 83(A6):692–697, 2005.

Bonvin D, Srinivasan B, Hunkeler D: Control and optimization of batch processes—Improvement of process operation in the production of specialty chemicals, *IEEE Cont Sys Mag* 26(6):34–45, 2006.

Boyd S, Vandenberghe L: *Convex optimization,* 2004, Cambridge University Press.

Bryson AE: *Dynamic optimization,* Menlo Park, CA, 1999, Addison-Wesley.

Bunin G, Wuillemin Z, François G, Nakajo A, Tsikonis L, Bonvin D: Experimental real-time optimization of a solid oxide fuel cell stack via constraint adaptation, *Energy* 39:54–62, 2012.

Chachuat B, Srinivasan B, Bonvin D: Adaptation strategies for real-time optimization, *Comp Chem Eng* 33(10):1557–1567, 2009.

Choudary BM, Lakshmi Kantam M, Lakshmi Shanti P: New and ecofriendly options for the production of speciality and fine chemicals, *Catal Today* 57:17–32, 2000.

Forbes JF, Marlin TE: Design cost: a systematic approach to technology selection for model-based real-time optimization systems, *Comp Chem Eng* 20:717–734, 1996.

Forbes JF, Marlin TE, MacGregor JF: Model adequacy requirements for optimizing plant operations, *Comp Chem Eng* 18(6):497–510, 1994.

Forsgren A, Gill PE, Wright MH: Interior-point methods for nonlinear optimization, *SIAM Rev* 44(4):525–597, 2002.

François G, Srinivasan B, Bonvin D, Hernandez Barajas J, Hunkeler D: Run-to-run adaptation of a semi-adiabatic policy for the optimization of an industrial batch polymerization process, *Ind Eng Chem Res* 43(23):7238–7242, 2004.

François G, Srinivasan B, Bonvin D: Use of measurements for enforcing the necessary conditions of optimality in the presence of constraints and uncertainty, *J Proc Cont* 15(6):701–712, 2005.

Gill PE, Murray W, Wright MH: *Practical optimization,* London, 1981, Academic Press.

Gisnas A, Srinivasan B, Bonvin D: Optimal grade transition for polyethylene reactors. In *Process Systems Engineering 2003, Kunming*, 2004, pp 463–468.

Marchetti A: *Modifier-adaptation methodology for real-time optimization.* PhD thesis Nr. 4449, EPFL, Lausanne, 2009.

Marchetti A, Amrhein M, Chachuat B, Bonvin D: Scale-up of batch processes via decentralized control. In *Int. Symp. on Advanced Control of Chemical Processes, Gramado,* 2006, pp 221–226.

Marchetti A, Chachuat B, Bonvin D: Modifier-adaptation methodology for real-time optimization, *Ind Eng Chem Res* 48:6022–6033, 2009.

Marlin T, Hrymak A: Real-time operations optimization of continuous processes, *AIChE Symp Ser* 93:156–164, 1997, CPC-V.

McAuley KB, MacGregor JF: On-line inference of polymer properties in an industrial polyethylene reactor, *AIChE J* 37(6):825–835, 1991.

McAuley KB, MacDonald DA, MacGregor JF: Effects of operating conditions on stability of Gas-phase polyethylene reactors, *AIChE J* 41(4):868–879, 1995.

Moore K: *Iterative learning control for deterministic systems, Advances in industrial control,* London, 1993, Springer-Verlag.

Rotava O, Zanin AC: Multivariable control and real-time optimization—An industrial practical view, *Hydrocarb Process* 84(6):61–71, 2005.

Skogestad S: Plantwide control: the search for the self-optimizing control structure, *J Proc Cont* 10:487–507, 2000.

Srinivasan B, Bonvin D: Dynamic optimization under uncertainty via NCO tracking: A solution model approach. In *BatchPro Symposium, Poros,* 2004, pp 17–35.

Srinivasan B, Bonvin D: Real-time optimization of batch processes via tracking the necessary conditions of optimality, *Ind Eng Chem Res* 46(2):492–504, 2007.

Srinivasan B, Primus CJ, Bonvin D, Ricker NL: Run-to-run optimization via control of generalized constraints, *Cont Eng Pract* 9(8):911–919, 2001.

Srinivasan B, Palanki S, Bonvin D: Dynamic optimization of batch processes: I. Characterization of the nominal solution, *Comp Chem Eng* 27:1–26, 2003.

Srinivasan B, Biegler LT, Bonvin D: Tracking the necessary conditions of optimality with changing set of active constraints using a barrier-penalty function, *Comp Chem Eng* 32(3):572–579, 2008.

Vassiliadis VS, Sargent RWH, Pantelides CC: Solution of a class of multistage dynamic optimization problems. 2. Problems with path constraints, *Ind Eng Chem Res* 33(9): 2123–2133, 1994.

Williams TJ, Otto RE: A generalized chemical processing model for the investigation of computer control, *AIEE Trans* 79:458, 1960.

Zhang Y, Monder D, Forbes JF: Real-time optimization under parametric uncertainty: A probabilistic constrained approach, *J Proc Cont* 12(3):373–389, 2002.

CHAPTER TWO

Incremental Identification of Distributed Parameter Systems[1]

Adel Mhamdi, Wolfgang Marquardt
Aachener Verfahrenstechnik - Process Systems Engineering, RWTH Aachen University, Aachen, Germany

Contents

Abstract

In this contribution, we present recent progress toward a systematic work process called model-based experimental analysis (MEXA) to derive valid mathematical models for kinetically controlled reaction and transport problems which govern the behavior of (bio-)chemical process systems. MEXA aims at useful models at minimal engineering effort. While mathematical models of kinetic phenomena can in principle be developed using standard statistical techniques including nonlinear regression and multimodel inference, this direct approach typically results in strongly nonlinear and large-scale mathematical programming problems, which may not only be computationally prohibitive but may also result in models which are not capturing the underlying

[1]This paper is based on previous reviews on the subject (Bardow and Marquardt, 2009; Marquardt, 2005) and reuses material published elsewhere (Marquardt, 2013).

Advances in Chemical Engineering, Volume 43
ISSN 0065-2377
http://dx.doi.org/10.1016/B978-0-12-396524-0.00002-7

physicochemical mechanisms appropriately. In contrast, incremental model identification, which is an integral part of the MEXA methodology, constitutes a physically motivated divide-and-conquer strategy to kinetic model identification.

1. INTRODUCTION

The primary subject of modeling is a (part of a) complete production process which converts raw materials in desired chemical products. Any process comprises a set of connected pieces of equipment (or process units), which are typically linked by material, energy and information flows. The overall behavior of the plant is governed by the behavior of its constituents and their nontrivial interactions. Each of these subsystems is governed by typically different types of kinetic phenomena, such as (bio-)chemical reactions or intra- and interphase mass, energy, and momentum transport. The resulting spatiotemporal behavior is often very complex and yet not well understood. This is particularly true if multiple, reactive phases (gas, liquid, or solid) are involved.

Mathematical models are in the core of methodologies for chemical engineering decisions (which) should be responsible for indicating how to plan, how to design, how to operate, and how to control any kind of unit operation (e.g., process unit), chemical and other production process and the chemical industries themselves (Takamatsu, 1983). Given the multitude of model-based engineering tasks, any modeling effort has to fulfill specific needs asking for different levels of detail and predictive capabilities of the resulting mathematical model. While modeling in the sciences aims at an understanding and explanation of observed system behavior in the first place, modeling in engineering is an integrated part of model-based problem solving strategies aiming at planning, designing, operating, or controlling (process) systems. There is not only a diversity of engineering tasks but also an enormous diversity of structures and phenomena governing (process) system behavior. Engineering problem solving is faced with such multiple dimensions of diversity. A kind of "model factory" has to be established in industrial modeling processes in order to reduce the cost of developing models of high quality which can be maintained across the plant life cycle (Marquardt et al., 2000).

Models of process systems are multiscale in nature. They span from the molecular level with short length and time scales to the global supply chain involving many productions plants, warehouses, and transportation systems. The major building block of a model representing some part of a process system (sometimes also called a balance envelope) is the differential balance equation, which is formulated for a selected set of extensive quantities (Bird et al., 2002). The balances constitute of hold-up, transport, and source terms which reflect the molecular behavior of matter on the continuum scale. Averaging is often applied to coarse grain the resolution of the model in time and space for complexity reduction (Slattery, 1999). Both, bridging from the molecular to the continuum scale by some kind of coarse-graining results unavoidably in so-called closure problems. Roughly speaking, a closure problem arises because the application of linear averaging operators to a nonlinear expression in a balance equation cannot be evaluated analytically to relate the average of such an expression to the averaged state variables (such as velocity, temperature, concentrations). The closure condition refers to some constitutive (in some cases even differential equation) model which relates the average of a nonlinear expression to the averaged state variables. A well-known closure problem refers to the determination of the Reynolds stress tensor which results from averaging the Navier–Stokes equations with respect to time (Pope, 2000). Even if such closure conditions are derived from theoretical considerations using some kind of scale-bridging approach, they typically require the identification of empirical parameters in the submodel structures or in extreme cases even the model structure (i.e., the mathematical expressions relating dependent and independent variables) itself. In particular, the so-called k-ε-model for the Reynolds stress tensor comprises a number of parameters which have to be determined from experiments (Bardow et al., 2008).

Since such model identification is a complex systems problem, a goal-oriented work process has to be established which systematically links high-resolution measurement techniques, mathematical modeling, real (laboratory), or virtual (simulation) experiments (typically on a finer scale) with the formulation and solution of so-called inverse problems (Kirsch, 1996). These inverse problems come in different flavors: they may be used to design the most informative experiment by fixing the experimental conditions in a given experimental setup appropriately (Pukelsheim, 2006; Walter and Pronzato, 1990), to estimate parameters (Bard, 1974; Schittkowski, 2002) in a given model structure or to discriminate among model structure candidates based on experimental evidence (Verheijen, 2003). Typically, the model identification task cannot be successfully tackled in one go. Rather, some kind of iterative refinement strategy is intuitively followed by the modeler to exploit the knowledge gained during the model development procedure. Probably the most important decision to be made is the level of detail to be included in the target model to result in a desired model resolution.

To this end, this contribution presents recent progress toward a systematic work process (Bardow and Marquardt, 2004a,b; Marquardt, 2005) to derive valid mathematical models for kinetically controlled reaction and transport problems which govern the behavior of (bio-)chemical process systems. Research on systematic work processes for mathematical model development, which combine experiments, data analysis, modeling, and model identification, dates at least back to the 1970s (Kittrell, 1970). However, the availability of current, more advanced experimental and theoretical techniques offer new opportunities to develop more comprehensive modeling strategies which are widely applicable to a variety of modeling problems. For example, a modeling process with a focus on optimal design of experiments has been reported by Asprey and Macchietto (2000).

Recently, the collaborative research center CRC 540, "Model-Based Experimental Analysis of Fluid Multi-Phase Reaction Systems" (cf. http://www.sfb540.rwth-aachen.de/), which was funded by the German Research Foundation (DFG), addressed the development of advanced modeling work processes comprehensively from 1999 to 2009. The research covered the development of novel high-resolution measurement techniques, efficient numerical methods for the solution of direct and inverse reaction and transport problems and the development of a novel, experimentally driven modeling strategy which relies on iterative model identification. This work process is called *model-based experimental analysis* (or MEXA for short) and aims at *useful models at minimal engineering effort.* While mathematical models of kinetic phenomena can in principle be developed using standard statistical techniques including nonlinear regression (Bard, 1974) and multimodel inference (Burnham and Anderson, 2002), this direct approach typically results in strongly nonlinear and large-scale mathematical programming problems (Biegler, 2010; Schittkowski, 2002), which may not only be computationally prohibitive but also result in models which are not capturing the underlying physicochemical mechanisms appropriately. In contrast, *incremental model identification* (or IMI for short), which is an integral part of the MEXA methodology, constitutes a physically motivated divide-and-conquer strategy to kinetic model identification.

IMI is not the first multistep approach to model identification. Similar ideas have been employed rather intuitively before in (bio-)chemical engineering. The sequence of flux estimation and parameter regression is, for example, commonly employed in reaction kinetics as the so-called differential method (Froment and Bischoff, 1990; Hosten, 1979; Kittrell, 1970). Markus et al. (1981) seem to be the first suggesting a simple version of

IMI to the identification of enzyme kinetics models. Bastin and Dochain (1990) have introduced model-free reaction flux estimation as part of a state estimation strategy with applications to bioreactors. More recently, a two-step approach has been applied for the hybrid modeling of fermentation processes (Tholudur and Ramirez, 1999; van Lith et al., 2002), where reaction fluxes are estimated first from measured data and neural networks or fuzzy models are employed to correlate the fluxes with the measurements. The crystal growth rate in mixed-suspension crystallization has been estimated directly from the population balance equations (Mahoney et al., 2002).

The idea has not only been around in the chemical engineering community. For example, Timmer et al. (2000) and Voss et al. (2003) use the two-step approach of flux estimation and rate law fitting in the modeling of nonlinear electrical circuits. Ramsay and coworkers used a similar method, called functional data analysis, in quantitative psychology to model lip motion (Ramsay et al., 1996) and handwriting (Ramsay, 2000), and in production planning (Ramsay and Ramsey, 2002). These diverse applications and our own experience lead us to the expectation, that IMI can be rolled out and tailored to many domains in engineering and the sciences.

This paper is structured as follows. Section 2 presents a general overview on standard approaches to model identification. IMI is introduced in Section 3. Sections 4 and 5 sketch the application of the IMI methodology exemplarily to challenging and relevant process modeling problems involving distributed parameter systems. They include multicomponent diffusion in liquids, (bio-)chemical reaction kinetics in single- and multiphase systems and energy transport in wavy falling film flows. The final Section 6 provides a summarizing discussion.

2. STANDARD APPROACHES TO MODEL IDENTIFICATION

In contrast to IMI (cf. Section 3), all established approaches to model identification neglect the inherent hierarchical structure of kinetic models of process systems (Marquardt, 1995). These so-called simultaneous model identification (SMI) approaches always assume that the model structure is correct and consider only the fully specified model. In particular, the decisions on the balance envelope and the desired spatiotemporal resolution, the selection of the models for the flux expression and the phenomenological coefficients are specified prior to adjusting the model response to the measured data by some kind of parameter estimation method. Since the submodels are typically not known, suitable model structures are selected by the modeler based on prior knowledge, experience, and intuition. Obviously, the complexity of the decision making process is enormous. The number of alternative model structures grows exponentially with the number of decision levels and the number of kinetic phenomena occurring simultaneously in the real system.

Any decision on a submodel will influence the predictive quality of the identified kinetic model. The model predictions are typically biased if the parameter estimation is based on a model containing structural error (Walter and Pronzato, 1997). The theoretically optimal properties of the maximum likelihood approach to parameter estimation (Bard, 1974) are lost, if structural model mismatch is present. More importantly, in case of biased predictions, it is difficult to identify which of the decisions on a certain submodel contributed most to the error observed.

One way to tackle these problems in SMI is the enumeration of all the combinations of the candidate submodel structures for each kinetic phenomenon. Such combinatorial aggregation inevitably results in a large number of model structures. The computational effort for parameter estimation grows very quickly and calls for high performance computing, even in case of spatially lumped models, to tackle the exhaustive search for the best model indicated by the maximum likelihood objective (Wahl et al., 2006). Even if such a brute force approach were adopted, initialization and convergence of the typically strongly nonlinear parameter estimation problems may be difficult since the (typically large number of) parameters of the overall model have to be estimated in one step (Cheng and Yuan, 1997). The lack of robustness of the computational methods may become prohibitive, in particular, in case of spatially distributed process models if they are nonlinear in the parameters (Karalashvili et al., 2011). Appropriate initial values can often not be found to result in reasonable convergence of an iterative parameter estimation algorithm.

After outlining the key ideas of the SMI methods, some discussion of the implementation requirements as a prerequisite for their roll-out in practical applications is presented next. The implementation of SMI is straightforward and can be based on a wealth of existing theoretical and computational tools. Implicitly, SMI assumes a *suitable experiment* and the *correct model structure* to be available. Then, the following steps have to be enacted:

SMI procedure

1. Make sure that all the model parameters are identifiable from the measurements (Quaiser et al., 2011; Walter and Pronzato, 1997). If necessary,

employ local identifiability methods (Vajda et al., 1989). If some parameters are not identifiable, the analysis could suggest which additional measurements are needed or how to reduce the model to make it identifiable. Select initial parameter values based on a priori knowledge and intuition.
2. Select conditions of initial experiment guided by statistical design of experiments (Mason et al., 2003).
3. Run the experiments for selected conditions to obtain experimental data.
4. Estimate the unknown parameters (Bard, 1974; Biegler, 2010; Schittkowski, 2002), most favorably by a maximum likelihood approach to get unbiased estimates, using the available experimental data.
5. Assess the confidence of the estimated parameters and the predictive quality of the model (Bard, 1974; Telen et al., 2012; Walter and Pronzato, 1997).
6. Design optimal experiments *for parameter precision* to improve the parameter estimates, reduce their variances, and thus improve the prediction quality of the model (Franceschini and Macchietto, 2008; Pukelsheim, 2006; Walter and Pronzato, 1990).
7. Reiterate the sequence of steps 3–5 until no improvement in parameter precision can be obtained.

If a set Σ of candidate model structures M_i has to be considered because the correct model structure is unknown, the SMI approach as outlined above cannot be applied without modification. We have to assume that the *correct model structure M_c is included in the set of candidate models*. Under this assumption, the above SMI procedure has to be modified as follows: Each of the tasks in steps 1, 4, and 5 have to be carried out sequentially for all the candidate models in the set Σ. A decision on the correct model in the set should not be based on the results of step 5, that is, the model with highest parameter confidence and the best predictive quality should not be selected, because the experiments carried out so far may not allow to distinguish between competing model candidates. An informed decision requires adding a step 6′ after step 6 has been carried out for each of the candidate models, the optimal design of experiments *for model discrimination* (Michalik et al., 2009a; Pukelsheim, 2006; Walter and Pronzato, 1990), to determine experiments which allow distinguishing between the models with highest confidence. The designed experiments are executed, the parameters in the (so far) most appropriate model structure are estimated. Since the optimal design of experiments relies on initial parameters which may be incorrect, steps 4 and 6′ have to be reiterated until the confidence in the most appropriate model structure in the candidate set cannot be improved and, hence, model M_c has been found. Once the model structure has been identified, steps 6 and 7 are performed to determine the best possible parameters in the correct model structure. The investigations should ideally only be terminated if the model cannot be falsified by any conceivable experiment (Popper, 1959).

A number of commercial or open-source tools (Balsa-Canto and Banga, 2010; Buzzi-Ferraris and Manenti, 2009) are available which can be readily applied to reasonably complex models, in particular to models consisting of algebraic or/and ordinary differential equations. Though this procedure is well established, a number of pitfalls may still occur (Buzzi-Ferraris and Manenti, 2009) which render the application of SMI a challenge even under the most favorable assumptions. An analysis of the literature on applications shows, that the identification of (bio-)chemical reaction kinetics has been of most interest to date.

Only little software support is available to the user for an optimal design of experiments for parameter precision (e.g., VPLAN, Körkel et al., 2004) and even less for model discrimination, which is required for a roll-out of the extended SMI procedure. Only few experimental studies have been reported which tackle model identification in the spirit of the extended SMI procedure.

3. INCREMENTAL MODEL IDENTIFICATION

IMI exploits the natural hierarchy in kinetic models of process systems. It relies on an incremental refinement of the model structure which is motivated by systematic model development as suggested by Marquardt (1995). Figure 2.1 shows schematically three model steps, which are denoted by model *B*, model *BF*, and model *BFR*, respectively. These steps and their relation to IMI are outlined in the following.

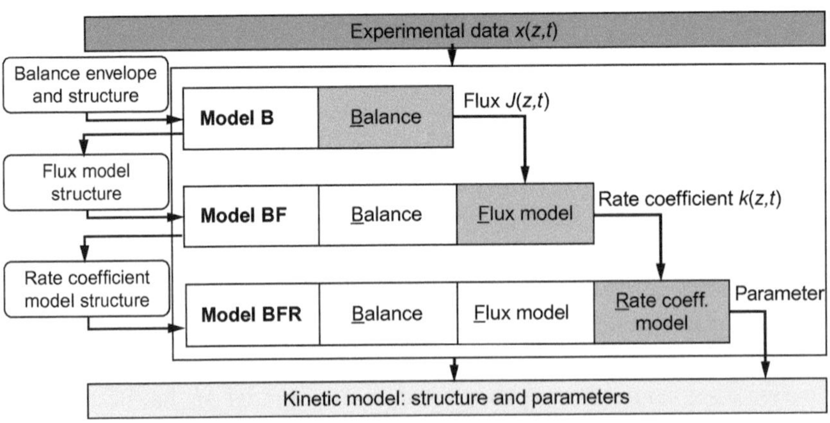

Figure 2.1 Incremental modeling and identification (Marquardt, 1995, 2005). (For color version of this figure, the reader is referred to the online version of this chapter.)

Model B. In model development, balance envelopes and their interactions are determined first to represent a certain part of the system of interest. The spatiotemporal resolution of the model is decided in each balance envelope, for example, the model may or may not describe the evolution of the behavior over time t and it may or may not resolve the spatial resolution in up to three space dimensions z. Quantities $y(z,t)$ such as mass, mass of a certain chemical species, energy, etc., are selected for which a balance equation is to be formulated. In the general case of spatiotemporally resolved models, the balance reads as

$$\begin{aligned} \frac{\partial y}{\partial t} &= -\nabla_z \cdot j_{t,y} + j_{s,y}, \quad z \in \Omega, t > t_0, \\ y(z,t_0) &= y_0(z), \\ \nabla_z y|_{\partial\Omega} &= j_{b,y}, \quad z \in \partial\Omega, \end{aligned} \qquad [2.1]$$

where $y(z,t)$ is propagated according to the transport term $j_{t,y}(z,t)$ and generated (or consumed) according to the source term $j_{s,y}(z,t)$ at any point in the interior of the balance envelope $\Omega \subset \mathbb{R}^n$, $n=1,2,3$. The symbol $j_{b,y}(z,t)$ refers to transport across the boundary $\partial\Omega$ of the balance envelope. Any quantity $y(z,t)$ is typically related to a set of measured quantities $x(z,t)$ by some constitutive relation

$$y = h(x). \qquad [2.2]$$

Note that no constitutive equations are considered yet to specify any of the terms $j_{f,y}, f \in \{t,s,b\}$, in Eq. (2.1) as a function of the intensive thermodynamic state variables x. While these constitutive equations are selected on the following decision level, the unknown terms $j_{f,y}$ are estimated in IMI directly from the balance equation. For this purpose, measurements of x with sufficient resolution in time t and/or space z are assumed. An unknown flux, $j_{f,y}$ can then be estimated from one of the balance equations (Eq. 2.1) as a function of time and/or space coordinates without specifying a constitutive equation.

Model BF. In model development, constitutive equations are specified for each term $j_{f,y}, f \in \{t,s,b\}$, in the balance equations (Eq. 2.1) on the next decision level. In particular,

$$j_{f,y}(z,t) = g_{f,y}\left(x, \nabla_z x, \ldots, k_{f,y}\right), \; f \in \{t,s,b\}. \qquad [2.3]$$

The symbols $k_{f,y}$ refer to some rate coefficient functions which depend on time and space. These constitutive equations could, for example, correlate interfacial fluxes or reaction rates with state variables x.

Similarly, in IMI, model candidates, as in Eq. (2.3), are selected or generated on decision level *BF* to relate the flux to rate coefficients, to measured states, and possibly to their derivatives. The estimates of the fluxes $j_{f,y}$ obtained on level *B* are now interpreted as *inferential* measurements. Together with the *real* measurements $x(z,t)$, one of these flux estimates can then be used to determine one of the rate coefficients $k_{f,y}$ as a function of time and space from the corresponding equation in Eq. (2.3), respectively. Often, the flux model can be analytically solved for the rate coefficient function $k_{f,y}$. These rate coefficient functions, for example, refer to heat or mass transfer or reaction rate coefficients.

Model BFR. In many cases, the rate coefficients $k_{f,y}(z,t)$ introduced in the correlations on level *BF* depend on the states $x(z,t)$ themselves. Therefore, a constitutive model

$$k_{f,y}(z,t) = r_{f,y}\left(x, \nabla_z x, \ldots, \theta_f\right), \; f \in \{t, s, b\}, \tag{2.4}$$

relating the rate coefficients to the states, has to be selected on yet another decision level named *BFR* (cf. Fig. 2.1).

Mirroring this last model development step in IMI, a model for the rate coefficients has to be identified. The model candidates, cf. Eq. (2.4), are assumed to only depend on the measured states, their spatial gradients, and on *constant* parameters $\theta_f \in \mathbb{R}^p$. If only a single candidate structure is considered, the parameters θ_f can be computed from the estimated functions $k_{f,y}(z,t)$ and the measured states $x(z,t)$ by solving a (typically nonlinear) algebraic regression problem. In general, however, a model discrimination problem has to be solved, where the most suitable model structure is determined from a set of candidates.

The cascaded decision making process in model development and model identification has been discussed for three levels which commonly occur in practice. However, model refinement can continue as long as the submodels of the last model refinement step not only involve constants θ_f, as in Eqs. (2.3) and (2.4), but rather coefficient functions which depend on state variables. While this is the decision of the modeler, it should be backed by experimental data and information deduced during incremental identification such as the confidence in the selected model structure and its parameters (Verheijen, 2003).

Error propagation is unavoidable within IMI, since any estimation error will clearly influence the estimation quality in the following steps. The resulting bias can, however, be easily removed by a final correction step, where a parameter estimation problem is solved for the best aggregated model(s) using very good initial parameter values. Convergence is typically achieved in one or very few iterations as experienced during the application of IMI to the challenging problems described in the following sections. Note that if no spatial resolution of the state variables is desired, the incremental approach for modeling and identification as introduced above does not change dramatically. Mainly, the dependence on the space coordinates z of the variables and Eqs. (2.1)–(2.4) is removed. All involved quantities will be a function of time only. In the following sections, we use capital letters to denote such quantities.

This structured modeling approach renders all the individual decisions completely transparent, that is, the modeler is in full control of the model refinement process. The most important decision relates to the choice of the model structures for the flux expressions and the rate coefficient functions in Eqs. (2.3) and (2.4). These continuum models do not necessarily have to be based on molecular principles. Rather, any mathematical correlation can be selected to fix the dependency of a flux or a rate coefficient as a function of intensive quantities. A formal, semiempirical but physically founded kinetic model may be chosen which at least to some extent reflects the molecular level phenomena. Examples include mass action kinetics in reaction modeling (Higham, 2008), Maxwell–Stefan theory of multicomponent diffusion (Taylor and Krishna, 1993) or established activity coefficient models like the Wilson, NRTL, or Uniquac models (Prausnitz et al., 2000). Alternatively, a purely mathematically motivated modeling approach could be used to correlate states with fluxes or rate coefficients in the sense of black-box modeling. Commonly used model structures include multivariate linear or polynomial models, neural networks, or vector machines among others (Hastie et al., 2003). This way, a *certain type of hybrid (or gray-box) model* (Agarwal, 1997; Oliveira, 2004; Psichogios and Ungar, 1992) arises in a natural way by combining first principles models fixed on previous decision levels with an empirical model on the current decision level (Kahrs and Marquardt, 2008; Kahrs et al., 2009; Romijn et al., 2008).

3.1. Implementation of IMI

Obviously, if the correct model structure is not known, it cannot be safely assumed that the correct model structure is part of the candidate set Σ; rather, the correct model, often comprising of a combination of many submodels, is not known. In this likely case, SMI should be replaced by IMI, the strength of which is to find an appropriate model structure composed of many submodels. The IMI procedure comprises the following steps:

IMI procedure

1. Develop *model B* (cf. Fig. 2.1): Decide on a balance envelope, on the desired spatiotemporal resolution and on the extensive quantities to be balanced, accounting for process understanding and modeling objectives.
2. Decide on the type of measurements necessary to estimate the unknown fluxes in *model B*.
3. Run informative experiments following, for example, a space-filling experiment design (Brendel and Marquardt, 2008), which aim at a balanced coverage of the space of experimental design variables. Note that model-based experiment design is not feasible, since an adequate model is not yet available.
4. Estimate the unknown fluxes $j_{f,y}(z,t)$ as a function of time and space coordinates using the measurements $x(z,t)$ and Eqs. (2.1)–(2.3). Use appropriate regularization techniques to control error amplification in the solution of this inverse problem (Engl et al., 1996; Huang, 2001; Reinsch, 1967), which are typically ill posed and thus very difficult to solve in a stable way, for example, without regularization, small errors in the data lead to large variations in the computed quantities.
5. Analyze the state/flux data and define a set of candidate flux models, Eqs. (2.3) and (2.4), with rate coefficient functions $k_{f,y}(z,t)$ parameterized in time and space. Fit the rate coefficient functions $k_{f,y}(z,t)$ of all candidate models to the state–flux data. Error-in-variables estimation (Britt and Luecke, 1975) should be used for favorable statistical properties, because both, the dependent fluxes as well as the measured states, are subject to error. A constant rate coefficient is obviously a reasonable special case of such a parameterization.
6. Form *candidate models* BF_i constituting balances and (all or only a few promising) candidate flux models. Reestimate the parameters in the rate coefficient functions $k_{f,y}(z,t)$ in all the *candidate models* BF_i to reduce the unavoidable bias due to error propagation (Bardow and Marquardt, 2004a; Karalashvili and Marquardt, 2010). Some kind of regularization of the estimation problem is required to enforce uniqueness of the estimation problem and to control error amplification in the estimates (Engl et al., 1996; Kirsch, 1996). Rank order the updated *candidate models* BF_i with respect to quality of fit using an appropriate statistical

measure such as Akaike's information criterion (AIC; Akaike, 1973; Burnham and Anderson, 2002) or a posteriori probabilities (Stewart et al., 1998). In case of constant rate coefficients, continue with step 8 replacing models *BFR* by *BF*.

7. Analyze the state/rate coefficient data and define a set of candidate rate coefficient models $r_{f,y}$, Eq. (2.4), for promising *candidate models* BF_i. Make sure that the parameters in the candidate rate coefficient models $r_{i,j}$ are identifiable from the state/rate coefficient data using identifiability analysis (Walter and Pronzato, 1997). Estimate the parameters $\theta_{i,j}$ in the rate coefficient models $r_{i,j}$ by means of an error-in-variables method (Britt and Luecke, 1975).
8. Form the candidate models $BFR_{i,j}$ by introducing the rate coefficient models $r_{i,j}$ in the models BF_i. Reestimate the parameters $\theta_{i,j}$ in the *candidate models* $BFR_{i,j}$ to remove the unavoidable bias due to error propagation.
9. Design *optimal experiments for model discrimination* using the set of *candidate models* $BFR_{i,j}$ to identify the most suitable model structure. Execute the design experiments and reestimate the parameters $\theta_{i,j}$ in the *candidate models* $BFR_{i,j}$ using the available experimental data. Reiterate this step until the confidence in the most suitable model structure BFR_c in the candidate set cannot be improved. If no satisfactory model structure can be identified in the set of candidate models, the set has to be revised by revisiting all previous steps.
10. Design *optimal experiments for parameter precision* using model BFR_c. Run the experiment and estimate the parameters θ_c in model BFR_c. Reiterate this step until the confidence in the parameters cannot be improved. If no satisfactory parameter confidence and prediction quality can be achieved, all previous steps have to be revisited.

Note that this IMI procedure as described above is not precise because its details depend on the type of model considered. The presented procedure is abstracted to roughly cover all types of models. How to adapt the procedure to each application area will be discussed below.

3.2. Ingredients for a successful implementation of IMI

A successful implementation of the incremental identification approach as discussed in Section 3 requires tailored ingredients:

- *high-resolution* (*in situ* and noninvasive) measurement techniques which provide field data of states like species concentrations, temperature, or velocities as a function of time and/or space coordinates,

- *algorithms* for model-free *flux estimation* by an inversion of the balance equations, a problem which is closely related to input estimation problems in systems and control engineering (Hirschorn, 1979) and to inverse problems (in particular, inverse source problems) in applied mathematics (Engl et al., 1996).
- *algorithms* for efficient *function estimation* comprising an (ideally error controlled) adaptive discretization of the unknown flux or rate coefficient functions in time and space coordinates (Brendel and Marquardt, 2009) and robust numerical methods for ill-conditioned, large-scale parameter estimation (Hanke, 1995).
- *methodologies* for the *generation*, *assessment*, and *selection of* the most suitable *model structures*; and
- model-based *methods* for the *optimal design of experiments* (Pukelsheim, 2006; Walter and Pronzato, 1990), which should be adapted to the requirements of IMI.

A detailed discussion of all these areas is definitely beyond the scope of this paper. Some aspects are, however, highlighted in the applications of IMI approach described in the following sections, where recent progress is exemplarily reported for selected kinetic modeling problems.

3.3. Application of IMI to challenging problems

The IMI has been developed and benchmarked with challenging problem classes dealing with the modeling of typical kinetic phenomena faced by chemical engineers during their activities in process design and operations, that is, reaction and multicomponent diffusive transport, transport and enzymatic reaction in gel particles, transport and reaction in dispersed liquid droplets, transport and reaction in liquid falling film. Obviously, we cannot address all the issues related to these systems in detail in this paper. Instead, we will focus on two problem classes: reaction–diffusion problems and flow systems with convective transport. Many publications already addressed special subproblems in both areas, where individual phenomena have been investigated based on the IMI procedure. The focus of this paper is devoted to the discussion of problems, where—in addition to the time dependence—we need to consider some spatial distributions of the unknown quantities.

However, we will start the discussion by considering the identification of reaction systems in a single homogeneous phase. This presentation of lumped parameter systems identification allows us to achieve a basic understanding of the IMI approach and a simple illustration of the methods needed to solve the identification problems in each step of IMI. A first step toward spatially extended distributed parameter systems refers to multiphase reactive systems where mass transport occurs in addition to chemical reaction. Diffusive mass transport requires the consideration of time and space dependences of the diffusion fluxes and hence the state variables. At the next level of complexity, we address falling liquid films and heat transfer during pool boiling, where the convective transport of mass or energy is involved. In all these cases, appropriate approaches must be developed to formulate the identification problems and efficiently deal with their solution and the very large amount of data.

We discuss in the following sections, some of the important issues related to the application of IMI for the following specific problem classes:

1. reaction–diffusion systems:
   - reaction kinetics in single- and multiphase systems,
   - multicomponent diffusion in liquids, and
   - diffusion in hydrogel beads.
2. systems with convective transport:
   - energy transport in falling liquid films and
   - pool boiling heat transfer.

These choices allow a gradual increase in the problem complexity and enable a clear assessment of the current state of knowledge for each specific problem and its associated class. In all cases, the experimental and computational aspects play an important role to allow for a successful application of the IMI approach.

4. REACTION–DIFFUSION SYSTEMS

4.1. Reaction kinetics

Mechanistic modeling of chemical reaction systems, comprising both, the identification of the most likely mechanism and the quantification of the kinetics, is one of the most relevant and still not yet fully satisfactorily solved tasks in process systems modeling (Berger et al., 2001). More recently, systems biology (Klipp et al., 2005) has revived this classical problem in chemical engineering to identify mechanisms, stoichiometry, and kinetics of metabolic and signal transduction pathways in living systems (Engl et al., 2009). Though this is the very same problem as in process systems modeling, it is more difficult to solve successfully because of three complicating facts: (i) there are severe restrictions to *in vivo* measurements of metabolite concentrations with sufficient (spatiotemporal) resolution, (ii) the numbers of metabolites and reaction steps are often very large, and (iii) the qualitative behavior of living systems changes with time giving rise to models with time-varying structure.

IMI has been elaborated in theoretical studies for a variety of reaction systems. Bardow and Marquardt (2004a,b) investigate the fundamental properties of IMI for a very simple reaction kinetic problem to elucidate error propagation and to suggest counteractions. Brendel et al. (2006) work out the IMI procedure for homogenous multireaction systems comprising any number of irreversible or reversible reactions. These authors investigate which measurements are required to achieve complete identifiability. They show that the method typically scales linearly with the number of reactions because of the decoupling of the identification of the reaction rate models. The method is validated with a realistic simulation study. The computational effort can be reduced by two orders of magnitude compared to an established SMI approach. Michalik et al. (2007) extend IMI to fluid multiphase reaction systems. These authors show for the first time, how the intrinsic reaction kinetics can be accessed without the usual masking effects due to interfacial mass transfer limitations. The method is illustrated with a simulated two-phase liquid–liquid reaction system of moderate complexity.

More recently, Amrhein et al. (2010) and Bhatt et al. (2010) have suggested an alternative decoupling method for single- and multiphase multireaction systems which is based on a linear transformation of the reactor model. The transformed model could be used for model identification in the spirit of the SMI procedure. Pros and cons of the decomposition approach of Brendel et al. (2006) and Michalik et al. (2007) and the one of Amrhein et al. (2010) and Bhatt et al. (2010) have been analyzed and documented by Bhatt et al. (2012). Selected features of IMI are elucidated for single- and multiphase reaction systems identification in the remainder of this section.

4.1.1 Single-phase reaction systems

Kinetic studies of reaction systems are often carried out in continuously or discontinuously operated stirred tank reactors or in differential flow-through reactors where the spatial dependency of concentrations and temperature can be safely neglected. Typically, the evolution of concentrations, temperatures, and flow rates is observed over time. Using the concentration data of a mixture of n_c chemical species, $C_i(t), i=1,\ldots,n_c$, the IMI procedure is instantiated for this particular case as follows. We refer to step n of the IMI procedure outlined in Section 3.1 by *IMI.n*.

4.1.1.1 Reaction flux estimation (*IMI.1–IMI.3*)

For homogeneous reactions in a single phase, the general material balances as given in Eqs. (2.1) and (2.2) specialize to result in *model B*, that is,

$$\frac{dN_i(t)}{dt} = Q(t)C_i^{in}(t) - Q(t)C_i(t) + F_i(t), \quad i = 1, \ldots, n_c, \qquad [2.5a]$$

$$N_i(t) = V(t)C_i(t), \qquad [2.5b]$$

where $N_i(t)$ denotes the mole number of chemical species i. The first two terms on the right hand side refer to the molar flow rates into and out of the reactor with known (or measured) molar flow rate $Q(t)$ and inlet concentrations $C_i^{in}(t)$. The last term in Eq. (2.5a) represents the unknown reaction flux of species i, that is, the molar amount of species i produced or consumed by all present chemical reactions. The measured concentrations $C_i(t)$ are converted into the *extensive* mole numbers $N_i(t)$ by multiplication with the known (or measured) reactor volume $V(t)$. Note that we tacitly assume measurements which are continuous in time to simplify the presentation. Obviously, real measurements are taken on a grid of discrete times. Hence, the equations may have to be interpreted accordingly.

All reaction fluxes $F_i(t)$ are unknown and have be estimated from the material balances using the measured concentration data $\widetilde{C}_i(t)$ for each species. Since the fluxes enter the balance Eq. (2.5a) linearly, the equations for each of the species are decoupled. Estimates of the fluxes $\hat{F}_i(t)$ may be computed individually by a suitable numerical approach. The flux estimation task is an ill-posed inverse problem, since we need to differentiate the concentration measurement data. This mainly means that small errors in the data will be amplified and thus lead to large variations in the computed quantities. However, this problem can successfully be solved by different regularization approaches, such as Tikhonov–Arsenin filtering (Mhamdi and Marquardt, 1999; Tikhonov and Arsenin, 1977) or smoothing splines (Bardow and Marquardt, 2004a; Huang, 2001).

Different methods are available for the choice of the regularization parameter, which is selected to balance data propagation and approximation (regularization) errors (Hansen, 1998). Two heuristic methods have been shown to give reliable estimates and are usually used if there is no a priori knowledge about the measurement error. The first method, generalized cross-validation (GCV), is derived from leave-one-out cross-validation where one concentration data point is dropped from the data set. The regularization parameter is chosen such that the estimated spline predicts the missing point best on average (Craven and Wahba, 1979; Golub et al.,

1979). The second method is the L-curve, which is a log–log–plot of a smoothing norm $\|\partial^2 C_i/\partial t^2\|$ over the residual norm $\|C_i - \tilde{C}_i\|$ (Hansen, 1998). This graph usually has a typical L-shape since the residual norm will be large for large λ, while the smoothing norm is minimized. For small λ, the residual norm will be minimized but the smoothing norm is large due to the ill-posed nature of the problem leading to oscillations in the solution. The optimal regularization parameter is therefore chosen as the point of the L-curve corresponding to the maximum curvature with respect to the regularization parameter. Computational routines for both methods are available (Hansen, 1999).

4.1.1.2 Reaction rate models (*IMI.4*)

The reaction fluxes refer to the total amount of a certain species produced or consumed in a reaction system. Since in a multireaction system, any chemical species i may participate in more than one reaction j, the reaction rates $R_j(t)$ have to be determined from the reaction fluxes $F_i(t)$, by solving the (usually nonsquare) linear system

$$F_i(t) = V(t)\sum_{j=1}^{n_R} v_{i,j} R_j(t), \quad i = 1, \ldots, n_c, \qquad [2.6]$$

using an appropriate numerical method. In Eq. (2.6), $v_{i,j}$ denotes the stoichiometric coefficient for the i-th species in the j-th reaction and n_R the number of reactions. The stoichiometric relations describing the reaction network may be cast into the $n_R \times n_c$ stoichiometric matrix $\boldsymbol{S} = [v_{i,j}]$. Thus, Eq. (2.6) may be written in vector form as

$$\boldsymbol{F}(t) = \boldsymbol{V}(t) S^T \boldsymbol{R}(t), \qquad [2.7]$$

where the symbol $\boldsymbol{F}(t)$ refers to the vector of n_c reaction fluxes, $\boldsymbol{R}(t)$ to the vector of reaction rates of the n_r reactions in the reaction system. Often the reaction stoichiometry is unknown; then, target factor analysis (TFA; Bonvin and Rippin, 1990) can be used to determine the number of relevant reactions and to test candidate stoichiometries suggested by chemical research. If more than one of the conjectured stoichiometric matrices is found to be consistent with the state/flux data, different estimates of $\boldsymbol{R}(t)$ are obtained in different scenarios to be followed in parallel in subsequent steps. The concentration/reaction-rate data are analyzed next to suggest a set Σ_j of candidate reaction rate laws (or purely mathematical relations) which relate each of the reaction rates R_j with the vector of concentrations $\boldsymbol{C}$ according to

$$R_j(t) = m_{j,l}\big(C(t), \theta_{j,l}\big), \quad j = 1, \ldots, n_R, l \in \Sigma_j. \qquad [2.8]$$

This model assumes isothermal and isobaric experiments, where the quantities $\theta_{j,l}$ are constants. A model selection and discrimination problem has to be solved subsequently for each of the reaction rates R_j based on the sets of model candidates Σ_j because the correct or at least best model structures are not known. These problems are, however, independent of each other. At first, the parameters $\theta_{j,l}$ in Eq. (2.8) are estimated from ($\hat{R}_j$ and $\hat{\boldsymbol{C}}$) data by means of nonlinear algebraic regression (Bard, 1974; Walter and Pronzato, 1997). Since the error level in the concentration data is generally much smaller than that in the estimated rates, a simple least-squares approach seems adequate. Thus, the parameter estimates result from

$$\hat{\theta}_{j,l} = \operatorname{argmin}\left\| \hat{R}_j(t) - m_{j,l}\big(\hat{C}(t), \theta_{j,l}\big) \right\|^2, \quad j = 1, \ldots, n_R, l \in \Sigma_j.$$

The quality of fit is evaluated by some means to assess whether the conjectured model structures (Eq. 2.8) fit the data sufficiently well.

4.1.1.3 Reducing the bias and ranking the reaction model candidates (*IMI.5*)

Equations (2.7) and (2.8) are now inserted into Eqs. (2.5a) and (2.5b) to form a complete reactor model. The parameters in the rate laws (Eq. 2.8) are now reestimated by a suitable dynamic parameter estimation method such as multiple shooting (Lohmann et al., 1992) or successive single shooting (Michalik et al., 2009d). Obviously, only the models of the sets Σ_j are considered, which have been identified to fit the data reasonably well. Very fast convergence is obtained, that is, often a single iteration is sufficient, because of the very good initial parameter estimates obtained from step *IMI.4*. This step reduces the bias in the parameter estimates computed in step *IMI.4* significantly. The model candidates can now be rank ordered, for example, by AIC (Akaike, 1973) for a first assessment of their relative predictive qualities.

4.1.1.4 Rate coefficient models (*IMI.6* and *IMI.7*)

In case of nonisothermal experiments, the quantities $\theta_{j,l}$ in the rate models (Eq. 2.8) are functions of temperature T. In this case, $\theta_{j,l}$ can be replaced by $k_{j,l}$, which has to be estimated first *without* specifying a rate coefficient model as in step *IMI.6*. Then, Eq. (2.8) is modified, and a parameterized rate coefficient model, such as the Arrhenius law,

$$k_{j,l} = \theta_{j,1} e^{\frac{\theta_{j,2}}{T}}$$
$$R_j(t) = k_{j,l} m_{j,l}(\mathbf{C}(t), \theta_{j,l}), \quad j = 1, \ldots, n_R, l \in \Sigma_j \qquad [2.9]$$

is introduced and the constant parameters $\theta_{j,1}$ and $\theta_{j,2}$ are estimated from the data $k_{j,l}(t)$ and $T(t)$ for every reaction j (see Brendel et al., 2006 for details).

4.1.1.5 Selection of best reaction model (*IMI.8* and *IMI.9*)

The identification of the reaction rate models may not immediately result in reliable model structures and parameters because of a lack of information content in the experimental data. Iterative improvement with optimally chosen experimental conditions should therefore be employed. Optimal experiments are designed first for model structure discrimination and then, after convergence, for parameter precision to yield the best model contained in the candidate sets.

4.1.1.6 Validation in simulation

To validate the IMI approach for identification reaction kinetics and investigate its properties and performance, the method has been investigated for many case studies in simulation. We illustrate the steps of the methodology for the acetoacetylation of pyrrole with diketene (see Brendel et al., 2006, for a more detailed discussion). By using simulated data, the results of the identification process can easily be compared to the model assumptions made for generating the data. The simulation is based on the experimental work of Ruppen (1994), who developed a kinetic model of the reaction system. In addition to the desired main reaction r_1 of diketene (D) and pyrrole (P) to 2-acetoacetyl pyrrole (PAA), there are three undesired side reactions r_2, r_3, r_4 that impair selectivity. These include the dimerization and oligomerization of diketene to dehydroacetic acid (DHA) and oligomers (OLs) as well as a consecutive reaction to the by-product G.

The reactions take place in an isothermal laboratory-scale semibatch reactor, to which a diluted solution of diketene is added continuously. The reactions r_1, r_2 and r_4 are catalyzed by pyridine (K), the concentration of which continuously decreases during the run due to addition of diluted diketene feed. Reaction r_3, which is assumed to be promoted by other intermediate products, is not catalyzed. A constant concentration of diketene in the feed C_D^{in} is assumed and zero for all other species. The initial conditions are known. The rate constant of the fourth reaction is set to zero, that is, this reaction is assumed not to occur in the network.

Using the assumed reaction rates and rate constants (Brendel et al., 2006), concentration trajectories are generated over a batch time $t_f = 60$ min. Concentration data are assumed to be available for the species D, PAA, DHA, OL, and G. Species P is assumed not to be measured. The measured concentrations are assumed to stem from a data-rich *in situ* measurement technique such as Raman spectroscopy, taken with the sampling period $t_s = 10$ s. Thus, a total of 361 data points for each species result. The data are corrupted with normally distributed white noise with standard deviations that differ for each species, depending on its calibration range.

In the first step, estimates of the reaction fluxes $F_i(t)$, $i = 1, \ldots, n_c$, are calculated using smoothing splines. A suitable regularization parameter is obtained by means of GCV. No reaction flux can be estimated for species P, since we assumed that it is not measured. Next, the stoichiometries of the reaction network have to be determined. The recursive TFA approach is applied to check the validity of the proposed stoichiometries and to identify the number of reactions occurring. The method successively accepts reactions r_2, r_1, and r_3 (in this order). Reaction r_4 does not take place in the simulation and is correctly not accepted. With this stoichiometric matrix, all reaction rates can be identified from the reaction fluxes present. The resulting time-variant reaction rates are depicted in Fig. 2.2 together with the true rates for comparison.

For the description of reaction kinetics, a set of model candidates for each accepted reaction is formulated as given in Table 2.1. To select a suitable model and compute the unknown model parameters, for each reaction, the available model candidates are fitted to the estimates of the concentrations and rates, both available as a function of time. For the first reaction, candidate 8 (cf. Table 2.1) can be best fitted to the estimated reaction rate and is identified as the most suitable kinetic law from the set of candidates.

Finally, for all three reactions the kinetics used for simulation as given in Table 2.1 were identified from the data available. The estimated rate constants $\hat{k}_1 = 0.0523$, $\hat{k}_2 = 0.1279$, and $\hat{k}_3 = 0.0281$ are very close to the values taken for simulation. The whole identification of the system using the proposed incremental procedure requires about 40 s on a standard PC (1.5 GHz).

For comparison, a simultaneous identification was applied to the data given, requiring dynamic parameter estimation for each combination of kinetic models and subsequent model discrimination. The simultaneous procedure correctly identifies the number of reactions and the corresponding kinetics. The reaction parameters are calculated as $\hat{k}_1 = 0.0532$, $\hat{k}_2 = 0.1281$, and

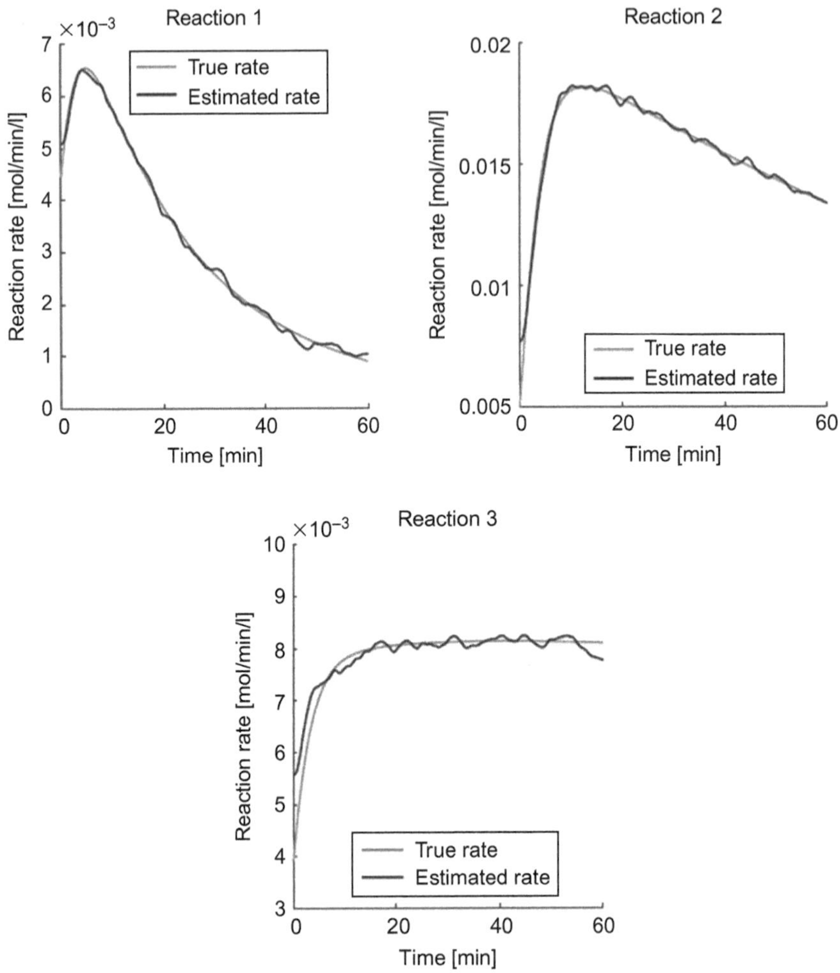

Figure 2.2 True and estimated reaction rates (Brendel et al., 2006).

$\hat{k}_3 = 0.028$, giving a slightly better fit compared to the incremental identification results. However, the computational cost is excessive; lying in the order of 34 h. Using IMI, an excellent approximation can be calculated in only a fraction of time.

4.1.1.7 Experimental validation

Recently, an experimental validation of IMI has been carried out (Michalik et al., 2007; Schmidt et al., 2009) for an enzymatic reaction, that is, the regeneration of NAD^+ to $NADH$, a cofactor used in many industrial

Table 2.1 Candidate models for all reactions

Reaction r_1 : $P + D \xrightarrow{K} PAA$	**Reaction r_2 :** $D + D \xrightarrow{K} DHA$	**Reaction r_3 :** $D \rightarrow OL$	**Reaction r_4 :** $PAA + D \xrightarrow{K} G$
$m_{1,1} = k_{1,1}$	$m_{2,1} = k_{2,1} C_D^2 C_K$	$m_{3,1} = k_{3,1} C_D$	$m_{4,1} = k_{4,1}$
$m_{1,2} = k_{1,2} C_D$	$m_{2,2} = k_{2,2} C_D$	$\mathbf{m_{3,2} = k_{3,2} C_D}$	$m_{4,2} = k_{4,2} C_D$
$m_{1,3} = k_{1,3} C_P$	$m_{2,3} = k_{2,3} C_D^2$	$m_{3,3} = k_{3,3} C_D^2$	$m_{4,3} = k_{4,3} C_{PAA}$
$m_{1,4} = k_{1,4} C_K$	$m_{2,4} = k_{2,4} C_D C_K$	$m_{3,4} = k_{3,4} C_D C_K$	$m_{4,4} = k_{4,4} C_K$
$m_{1,5} = k_{1,5} C_P C_D$	$\mathbf{m_{2,5} = k_{2,5} C_D^2 C_K}$	$m_{3,5} = k_{3,5} C_D^2 C_K$	$m_{4,5} = k_{4,5} C_{PAA} C_D$
$m_{1,6} = k_{1,6} C_P C_K$	$m_{2,6} = k_{2,4} C_K$	$m_{3,6} = k_{3,6} C_K$	$m_{4,6} = k_{4,6} C_{PAA} C_K$
$m_{1,7} = k_{1,7} C_D C_K$			$m_{4,7} = k_{4,7} C_D C_K$
$\mathbf{m_{1,8} = k_{1,8} C_P C_D C_K}$			$\mathbf{m_{4,8} = k_{4,8} C_{PAA} C_D C_K}$
$m_{1,9} = k_{1,9} C_D C_P^2$			$m_{4,9} = k_{4,9} C_D C_{PAA}^2$
$m_{1,10} = k_{1,10} C_D^2 C_P$			$m_{4,10} = k_{4,10} C_D^2 C_{PAA}$

The assumed true models are indicated in bold face (Brendel et al., 2006).

enzymatic reactions where it is reduced to NAD^+. The reaction takes place in aqueous solution using formic acid as a proton donor. There are two reactions of interest, the reversible regeneration reaction which forms $NADH$ and CO_2 as a by-product, and an undesired irreversible decomposition of the product $NADH$. The experiments were carried out in a micro-cuvette reactor of 300 μl, where the NADH concentration was measured with high accuracy and high resolution using UV/Vis spectroscopy at an excitation wavelength of 340 nm. The application of IMI to this industrially relevant problem (Michalik et al., 2007) resulted in a reaction kinetic model with much better predictive quality compared to existing and widely used literature models (Schmidt et al., 2009).

4.1.2 Multiphase reaction systems

The application of IMI to multiphase reactions is of great practical interest, because it is extremely difficult to access the intrinsic kinetics of a chemical reaction which is completely independent of mass transfer effects. Current practice in kinetic modeling of two-phase systems aims at experimental conditions where the chemical reaction is clearly rate limiting and the effect of the (very fast) mass transfer between the phases can be safely neglected. Obviously, this strategy is quite restrictive and inevitably results in systematic errors in reaction kinetics due to mass transfer contributions. IMI can remedy this long-standing problem in a straightforward manner as shown by Michalik et al. (2009a,b,c,d).

Let us assume isothermal experiments in a stirred tank reactor which is operated in batch mode (e.g., no material is exchanged with the environment) at isothermal conditions. A liquid–liquid (or liquid–gas) reaction is carried out, where the reaction occurs in one of the phases, say φ_a, only. The experiment is set up such that two well-mixed segregated phases φ_a and φ_b occur where spatial dependencies of the state variables are negligible. This assumption can easily be implemented by means of appropriate mixing and stabilization of the interface. Concentrations $C_i^a(t)$ and $C_i^b(t)$ of the relevant species $i=1,\ldots,n_c$ are assumed to be measured (e.g., by some kind of optical spectroscopy) in both phases. The material balances, specializing the general equations (Eq. 2.1) for species $i=1,\ldots,n_c$, read as

$$\begin{aligned} V^a \frac{\mathrm{d}C_i^a(t)}{\mathrm{d}t} &= J_i(t) + F_i(t), \\ V^b \frac{\mathrm{d}C_i^b(t)}{\mathrm{d}t} &= -J_i(t). \end{aligned} \qquad [2.10]$$

The volumes V^a and V^b of both phases are assumed constant and known for the sake of simplicity. The symbols $J_i(t)$ and $F_i(t)$ refer to the mass transfer rate of species from phase φ_b to phase φ_a and the reaction flux in phase φ_a, respectively.

Steps *IMI.1* to *IMI.3* have to be slightly modified compared to the case of homogenous reaction systems discussed in Section 5.1. In particular, the balance of phase φ_b and the measurements of the concentrations $C_i^b(t)$ are used to estimate the mass transfer rates $J_i(t)$ first without specifying a mass transfer model. These estimated functions can be inserted into the balances of phase φ_a to estimate the reaction fluxes $F_i(t)$ without specifying any reaction rate model. The intrinsic reaction kinetics can easily be identified in the subsequent steps *IMI.4* to *IMI.9* from the concentration measurements $C_i^a(t)$ and estimates of the reaction fluxes $F_i(t)$. Obviously, mass transfer models can be identified in the same manner if the mass transfer rates and the concentration measurements in both phases $C_i^a(t)$ and $C_i^b(t)$ are used accordingly.

4.1.2.1 Experimental validation

The basic idea of IMI of multiphase reaction systems has been evaluated in a simulated case study of a fluid two-phase system by Michalik et al. (2009a,b,c,d). These authors show that the intrinsic reaction kinetics can indeed be identified at high precision. Kerimoglu et al. (2011, 2012) validated Michalik's method for the first time in a real experimental study of a multiphase system. The chemical system studied comprises a Friedel–Crafts acylation of anisole. It follows a complex catalytic reaction mechanism with two reactants and two products. Several reaction rate models, both elementary and complex, were analyzed. The quality of the candidate models has been assessed by the residual sum of squares serving as an objective function and the AIC. Optimal experiments were designed to improve model quality using the AWDC criterion (Michalik et al., 2009a). It was found out that a reaction rate model comprising only two rate constants for the forward and backward reactions respectively fits best with a small confidence interval in contrast to a mechanism suggested in literature before. Since, mass transfer and chemical reaction can be systematically decoupled in the identification procedure, the best fitting mass transfer model of the four species involved can also be determined from the same experimental data set. Several mass transfer models of increasing complexity were tested. The results show that a simple model which neglecting diffusion cross-effects fits the experimental data best. An optimal design of experiments is currently being conducted to improve the reliability of the kinetic models.

4.2. Multicomponent diffusion in liquids

Despite extensive and lasting research efforts on diffusive transport, there is still a surprising lack of experimentally validated diffusion models, in particular for complex multicomponent liquid mixtures (Bird, 2004). This is in stark contrast to the relevance of the quantitative representation of diffusion to support the design of technical equipment. For example, the interplay of multicomponent diffusion and chemical reaction determines the selectivity toward the desired product in industrial reactors. In particular, in microreactors where mixing is only due to diffusion because of the laminar flow conditions, the complex mixing and diffusion patterns are decisive for reactor performance (Bothe et al., 2010).

The application of IMI to diffusive mass transport in liquid systems as introduced by Bardow et al. (2003, 2006) is featured in this section. It is based on a recently introduced Raman diffusion experiment, where the interdiffusion of two initially layered liquid mixtures is observed by Raman spectroscopy under isothermal conditions. Raman spectra of all species are measured on a line in the axis of a tailored cuvette at high resolution in time and space (cf. Fig. 2.3). The molar concentrations $c_i(z,t)$ of all species i are

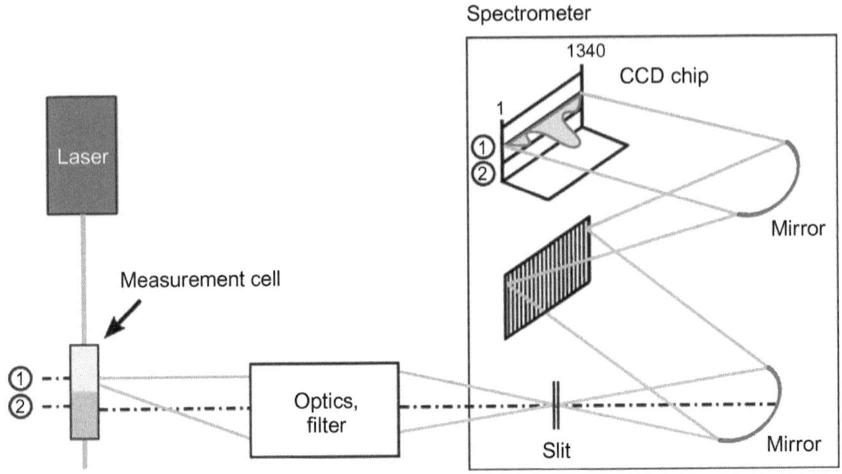

Figure 2.3 Experimental setup of 1D-Raman spectroscopy for diffusivity measurements (Kriesten et al., 2009).

determined from the Raman spectra by means of indirect hard modeling (Alsmeyer et al., 2004; Kriesten et al., 2008) at high accuracy. Figure 2.4 shows exemplarily concentration profiles as a function of space and time in a chemically homogeneous binary system consisting of cyclohexane and ethyl acetate obtained during such a diffusion experiment (Kriesten et al., 2009).

Using the concentration data of a mixture of n_c species, $c_i(z,t)$, $i=1,\ldots,n_c$, the IMI procedure is instantiated for this particular case as follows.

4.2.1 Estimationof diffusive fluxes (IMI.1–IMI.4)

The diffusion process is assumed to be well described by a spatially one-dimensional (1D) model, that is, $z \in [0, L]$, where L is the length of the vertical diffusion cell starting at its bottom (cf. Fig. 2.3). The adaption of the general balance equation (Eq. 2.1) results in *model B*, that is, a system of mass balance equations for all species $i=1,\ldots,n_c$:

$$\begin{aligned} &\frac{\partial c_i(z,t)}{\partial t} = -\frac{\partial j_i(z,t)}{\partial z}, \quad z \in [0,L], t > t_0, \, i=1,\ldots,n_c-1, \\ &c_i(z,t_0) = c_{i,0}(z), \\ &\quad j_i(0,t) = j_i(L,t) = 0. \end{aligned} \tag{2.11}$$

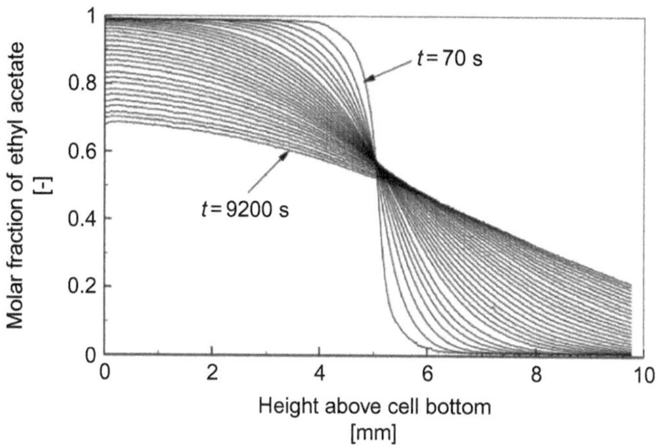

Figure 2.4 Space- and time-dependent concentration profiles of ethyl acetate during a diffusion experiment (Kriesten et al., 2009).

The diffusive fluxes $j_i(z,t)$ are defined relative to the volume average velocity, which is usually negligible (Tyrell and Harris, 1984). Other reference frames for diffusion are clearly possible (cf. Taylor and Krishna, 1993). However, the choice of the laboratory reference frame is especially convenient in experimental studies. The $n_c - 1$ independent diffusive fluxes $j_i(z,t)$ are unknown and have to inferred by an inversion of each of the evolution equations (Eq. 2.11) using measured concentration profiles $\widetilde{c}_i(z_m, t_m)$ at positions z_m and times t_m. Clearly, the choice of the measurement positions and times influences the estimation of the diffusive fluxes. Optimal values may be found using experiment design techniques (Bardow, 2004). By integrating Eq. (2.11), we obtain

$$j_i(z,t) = \int_0^z \frac{\partial \widetilde{c}_i(\zeta, t)}{\partial t} \mathrm{d}\zeta, \quad z \in [0, L], t > t_0, i = 1, \ldots, n_c - 1. \qquad [2.12]$$

To render the diffusive fluxes $j_i(z,t)$ without specifying a diffusion model, the measurements have to be differentiated with respect to time t first and the result has to be integrated over the spatial coordinate next. There is only a linear increase in computational complexity due to the natural decoupling of the multicomponent material balances (Eq. 2.11). An extended Simpson's rule is used here to evaluate the integral. The main difficulty in the evaluation of Eq. (2.12) though is the estimation of the time derivative of the measured concentration data. This is known to be an ill-posed problem, that is, small errors in the data will be amplified (Hansen, 1998). Therefore, smoothing splines regularization (Reinsch, 1967) are used, where the time derivatives are computed from a smoothed approximation of the data $\tilde{c}_i$. This method has successfully been applied for binary and ternary diffusion problems (Bardow et al., 2003, 2006). A smoothed concentration profile $\hat{c}_i$ is the solution of the minimization problem

$$\min_{c_i} \|c_i - \tilde{c}_i\| + \lambda \left\| \frac{\partial^2 c_i}{\partial t^2} \right\|. \qquad [2.13]$$

This approach corresponds to the well-known Tikhonov regularization method (Engl et al., 1996). λ is the regularization parameter, which is selected to balance data propagation and approximation (regularization) errors.

It should be noted that the estimation of a diffusive flux requires only the solution of the linear problem, Eq. (2.13), independent of the number of candidate models. All following estimation problems on the flux and coefficient model level (Fig. 2.1) are only algebraic. This decoupling of the problem reduces the computational expense substantially. But the decoupling comes at the price of an infinite-dimensional estimation problem of the molar flux, which is only feasible given sufficient data.

4.2.2 Diffusion flux models (IMI.5)

One or more flux models have to be introduced next. The generalized Fick model (or the Maxwell–Stefan model which is not further considered here) is a suitable choice. In case of binary mixtures, the Fick diffusion coefficient $D_{1,2}(z,t)$ can be determined at any point in time and space by solving the flux equation

$$j_1(z,t) = -D_{1,2}(z,t)\frac{\partial c_1(z,t)}{\partial z}, \qquad [2.14]$$

using the estimates $\hat{j}_1(z,t)$ and $\hat{c}_1(z,t)$ as data, which have already been computed in the previous step.

This strategy does not carry over directly to multicomponent mixtures because the diffusive flux is a linear combination of all concentration gradients:

$$j_n(z,t) = -\sum_{m=1}^{n_c-1} D_{n,m}(z,t)\frac{\partial c_m(z,t)}{\partial z}, \quad n = 1,\ldots,n_c - 1. \qquad [2.15]$$

Rather, the $n_c - 1$ diffusion coefficients have to be parameterized somehow. For example, some approximating spatiotemporal function could be chosen to formulate a least-squares problem which determines the diffusion coefficients $D_{n,m}(z,t)$ as function of time and space coordinates. Alternatively, a physically based parameterization (e.g., a diffusion coefficient model) could be chosen to lump *IMI.4* and *IMI.6* and eliminate *IMI.5*.

4.2.3 *Reducing the bias* (IMI.6)

The *model BF* can be formed by introducing Eq. (2.14) into Eq. (2.13). The diffusion coefficient functions can be reestimated using the results of the previous step as initial values of the parameter estimation problem to reduce the bias due to error propagation.

4.2.4 *Diffusion coefficient models* (IMI.7 *and* IMI.8)

To correlate the estimated diffusion coefficient data $D_{n,m}(z,t)$ with the measured concentrations, diffusion coefficient models can now be chosen:

$$D_{n,m} = r_{n,m,l}(c,\theta_l), \quad n,m = 1,\ldots,n_c - 1, l \in \Sigma_{n,m}. \qquad [2.16]$$

A model selection problem has to be solved. The parameters θ_l are identified by error-in-variables estimation (Britt and Luecke, 1975). The bias can be removed by inserting Eq. (2.16) into Eq. (2.15) and the result into Eq. (2.11) and reestimating the parameters. The models can be ranked with respect to model quality by some statistical measure (Burnham and Anderson, 2002; Stewart et al., 1998).

To be specific, we consider a binary diffusion and the case where no reasonable model candidate can be formulated. Therefore, a general parameterization for the diffusion coefficient $\hat{D}_{1,2}$ is introduced. The parameterization should be capable of approximating any function. Hanke and Scherzer (1999) suggested to divide the concentration range into p intervals X_k, $k = 1,\ldots,p$. The diffusion coefficient is approximated by a piece-wise constant function in each interval, that is, $\hat{D}_{1,2}(z,t) = \theta_k$ for $c \in X_k$. By collecting the estimated diffusion coefficients in a vector $\hat{\boldsymbol{D}}_{1,2}$ and the parameters $\boldsymbol{\theta} = [\theta_1,\theta_2,\ldots,\theta_p]^T$, we get the residual equations

$$\hat{\boldsymbol{D}}_{1,2} = \boldsymbol{A}\hat{\boldsymbol{\theta}}. \qquad [2.17]$$

The matrix $\boldsymbol{A}$ is extremely sparse containing only a single 1 per row denoting the appropriate concentration level. It turns out in practice that it is more advantageous to insert the diffusion coefficient model into the transport law (Eq. 2.14) to avoid explicit division by the spatial concentration gradient. The resulting residual equations read

$$\hat{\boldsymbol{J}}_1 = \overline{\boldsymbol{A}}\hat{\boldsymbol{\theta}} \qquad [2.18]$$

where $\overline{\boldsymbol{A}}$ contains the estimated spatial derivatives of the concentrations and $\hat{\boldsymbol{J}}_1$ the estimated diffusive fluxes, both sampled at the measured time instants and space positions. The estimation problem for the unknown parameter vector $\boldsymbol{\theta}$ may be stated as a least-squares estimation problem, for example,

$$\hat{\boldsymbol{\theta}} = \mathrm{argmin}_{\theta}\left\|\hat{\boldsymbol{J}}_1 - \overline{\boldsymbol{A}}\boldsymbol{\theta}\right\|. \qquad [2.19]$$

For the solution of such discrete ill-posed problems, several methods have been proposed (Hansen, 1998). Because of the large problem size and the sparsity of $\overline{\boldsymbol{A}}$, iterative regularization methods are the most appropriate choice (Hanke and Scherzer, 1999). This procedure leads to an unstructured model for the unknown diffusion coefficient. It is represented as a piecewise constant function of concentration.

4.2.5 Selecting the best diffusion model (IMI.9 and IMI.10)

The possible lack of information content in the experimental data can be remedied by an iterative improvement with optimally chosen experimental conditions to finally yield the best diffusion model.

4.2.6 Validation in simulation

In order to assess the IMI approach for diffusive mass transfer, we summarize the simulated case study of Bardow et al. (2003). This allows us to evaluate the different steps of the incremental algorithm. The "true" relation between the binary diffusion coefficient and concentration is assumed as

$$D_{1,2} = \vartheta_1 + \vartheta_2(x_1 - 0.5)^2 + \vartheta_3(x_1 - 0.5)^6. \qquad [2.20]$$

This constitutive equation should be recovered from measurements of the molar fraction x_1. The example considered is particularly challenging because of the nonmonotonous behavior of the diffusion coefficient (Cannon and DuChateau, 1980). To generate the data, the diffusion cell is assumed to be of length $L = 10$ mm. At $t = 0$, the lower half is filled with pure component 1, pure component 2 is layered on top. Measurements of the mole fraction $\tilde{x}_1(z_m, t_m)$ are taken with a resolution of 0 $\Delta z = 0.1$ mm and $\Delta t = 120$ s. The experiment runs for 2 h. Gaussian noise with a level of $\sigma = 0.01$ has been added to the simulated mole fraction data. This corresponds to very unfavorable experimental conditions for binary Raman experiments (Bardow et al., 2003).

To apply IMI, the concentrations $\tilde{c}_1(z_m, t_m)$ need to be computed from the mole fractions $\tilde{x}_1(z_m, t_m)$. A piecewise constant representation of the diffusion coefficient $\hat{D}_{1,2}$ is estimated using the computed flux values by solving the optimization problem (Eq. 2.19). Here, the conjugate gradient (CG) method is employed using the Regularization Toolbox (Hansen, 1999). A preconditioner enhancing smoothness may be used. The number of CG-iterations serves as the regularization parameter. It is chosen by the L-curve as shown in Fig. 2.5. The smoothing norm here approximates the second derivative of $D_{1,2}$ with respect to concentration; the residual norm is the objective function value.

The estimated and the true concentration dependence of the diffusion coefficient are compared in Fig. 2.6. The shape of the concentration dependence is well captured. It should be noted that only data from *one* experiment were used. Commonly, more than 10 experiments are employed (Tyrell and Harris, 1984). Nevertheless, the error is well below 5% for most of the concentration range. The minima and the maximum are found quite accurately in location and value. The values of the diffusion coefficient at the boundaries of the concentration range are not identifiable since the measured concentration gradient vanishes there. Better estimates are only possible with a

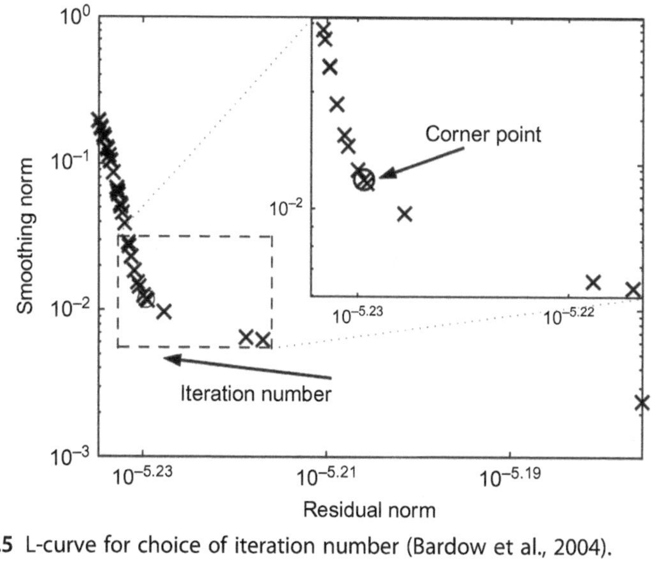

Figure 2.5 L-curve for choice of iteration number (Bardow et al., 2004).

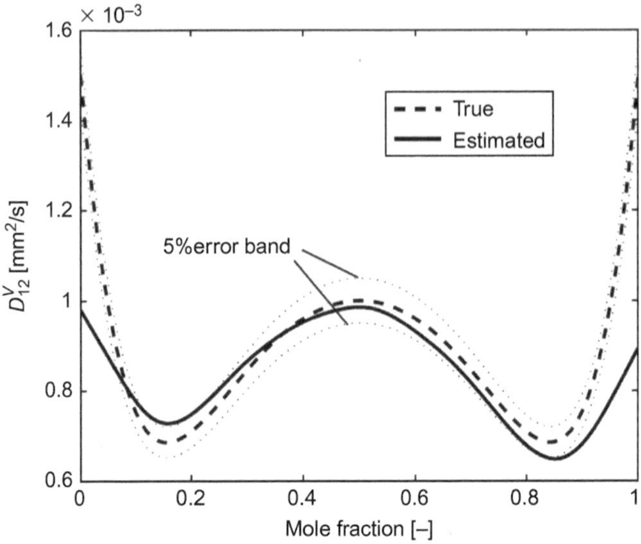

Figure 2.6 Estimated and true diffusion coefficient as a function of molar fraction (Bardow et al., 2004).

more sophisticated experimental procedure which establishes large gradients in these regions of dilution, for example by some kind of periodic forcing at the boundaries. The discretization level of the diffusion coefficient had only minor influence on the final result. Here, the concentration range was split into 500 intervals, that is, 500 parameters have to be estimated (cf. Eq. 2.19). This clearly prohibits the use of SMI whereas IMI takes an average CPU time of only 8 s on a standard desktop PC. This substantial reduction in computational time is mainly due to the decoupling of the problem. The use of an equation error scheme further reduces computational cost because the repeated solution of the model is avoided.

4.2.7 Experimental validation

The presented strategy has been validated in a number of experimental studies including the determination of binary and ternary Fick diffusion coefficients with a very low number of Raman experiments (Bardow et al., 2003, 2006) and the identification of the full concentration dependency of the binary Fick diffusion coefficient by means of a single Raman interdiffusion experiment (Bardow et al., 2005) and two additional NMR self-diffusion experiments at infinite dilution to improve accuracy (Kriesten et al., 2009).

4.3. Diffusion in hydrogel beads

In this section, we briefly discuss a more challenging, identification task, where reaction and diffusion occur simultaneously in an enzyme-catalyzed reaction in a hydrogel carrier. Enzyme catalyzed reactions constitute an efficient alternative for the production of various chemicals, drugs, materials, and fuels. However, several drawbacks complicate their application in large-scale industrial processes. An approach to overcome these difficulties is by immobilizing the enzymes, for instance in hydrogel beads, which are suspended in a solvent bulk phase as depicted in Fig. 2.7 (Ansorge-Schumacher et al., 2006). Moreover, enzyme immobilization facilitates downstream processing and reduces the overall process cost because the enzyme immobilizates can easily be recovered and reused.

The rational design of enzyme immobilizates is, however, more complex than that of homogeneous systems since mass transfer and diffusion can become rate limiting (Bauer et al., 2002; Berendsen et al., 2006; Halling, 1987). Moreover, diffusion and mass transfer have to be modeled in addition to the reaction. To ease the model identification process, it is usually assumed that the kinetic parameters of immobilized enzymes are identical to those of enzymes in solution (van Roon et al., 2006). Nevertheless, immobilization of enzymes can affect their kinetic constants, as observed so far for covalent binding techniques (Berendsen et al., 2006; Buchholz, 1989). Since it is yet unknown, whether immobilization of enzymes in hydrogel beads also alters reaction kinetic constants, recent research work (Michalik et al., 2007; Zavrel et al., 2010) has been addressing potential impact of immobilization on enzyme kinetics. This work has demonstrated that such complex systems can only be identified following a systematic

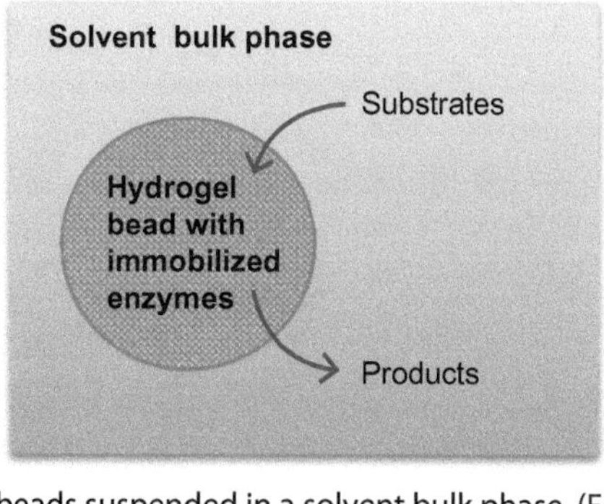

Figure 2.7 Hydrogel beads suspended in a solvent bulk phase. (For color version of this figure, the reader is referred to the online version of this chapter.)

process using spatially and temporally resolved measurement data stemming from optimally designed experiments.

Identification of the reactive biphasic hydrogel system shown in Fig. 2.7 has to consider three simultaneously occurring kinetic phenomena, that is, (i) enzyme reaction, (ii) mass transfer across the phase interface, and (iii) diffusion within the hydrogel bead.

Modeling assumes the organic (bulk) phase to be well mixed such that spatial dependencies of the state variables are negligible. In the hydrogel bead, we assume the variables to depend on the radial position z only. For each species, $i=1,\ldots,n_c$ we denote its concentration by $C_i^a(t)$ in the bulk phase and $c_i^b(z,t)$ in the bead. Let V^a be the bulk volume and A^b the surface of the bead. We denote by $j_i^b(z,t)$ the molar diffusive flux of species i and $f_i^b(z,t)$ the reaction flux of the only macro-kinetic reaction occurring inside the bead.

The material balances for the bulk and the bead, specializing the general equations (Eq. 2.1) for species $i=1,\ldots,n_c$, read as follows:

$$\begin{aligned} \frac{\partial c_i^b(z,t)}{\partial t} &= -\frac{1}{z^2}\left[\frac{\partial}{\partial z}\left(z^2 j_i^b(z,t)\right)\right] + f_i(z,t), \\ V^a \frac{\mathrm{d}C_i^a(t)}{\mathrm{d}t} &= -A^b j_i^b(z_b,t), \end{aligned} \qquad [2.21]$$

where z_b is the radius of the bead. The independent diffusive fluxes $j_i^b(z,t)$ and the reaction fluxes $f_i(z,t)$ are unknown and have to be inferred from measured concentration profiles $\tilde{c}_i^b(z,t)$ and $\tilde{C}_i^a(t)$ in both (bead and bulk) phases. Once these reaction and mass transfer flux estimates are available, they can be used as data for the next steps of the IMI procedure. It is, however, obvious that the system is not identifiable since the fluxes $j_i^b(z,t)$ and $f_i(z,t)$ cannot be estimated simultaneously, even if all concentration fields were observed. Therefore, the identification of the complete system is not possible in a single step.

To allow for a sound identification of the complex reaction–diffusion system, we may first investigate simpler system configurations with only a single kinetic phenomena occurring, and gather in a second step the available information to identify the complete system. This procedure has two advantages. Firstly, good initial guesses for the parameter estimation of the more complex models are obtained by the identification of the less complex models, and, secondly, potential interactions of the kinetic phenomena as well as a potential effect of the reaction systems on the kinetics are identified this way.

For instance, the reaction kinetics may be identified first in an experiment involving a homogeneous, ideally mixed reaction system, where the enzyme is dissolved in aqueous solution. The resulting reaction fluxes $f_i(z,t)$ could then be introduced into Eq. (2.21) to infer the diffusive fluxes in a similar way as described in Section 4.2. This strategy has been investigated by Michalik et al. (2007).

However, the enzyme kinetics might be influenced by immobilization (Berendsen et al., 2006; Buchholz, 1989). To investigate this influence, the diffusive flux $j_i^b(z,t)$ could be pragmatically modeled by Fick's law with effective diffusion coefficients D_i. Equation (2.21) can then be rewritten as

$$\frac{\partial c_i^b(z,t)}{\partial t} = \frac{1}{z^2}\left[\frac{\partial}{\partial z}\left(z^2 D_i \frac{\partial}{\partial z} c_i^b(z,t)\right)\right] + f_i(z,t). \qquad [2.22]$$

In this system, the reaction flux $f_i(z,t)$ may be inferred from measured concentration profiles $\tilde{c}_i^b(z,t)$. Two-photon confocal laser scanning microscopy (CLSM) maybe applied as measuring technique, since this allows access to concentration data at any radial position in the hydrogel bead. A sample measurement is shown in Fig. 2.8 (Schwendt et al., 2010). The remaining steps of the IMI are carried out according to the same procedure as in concentrated systems (cf. Section 4.1). However, there are some complications which have not been faced in the other types of problems. First, the second derivative of the concentration measurement data with respect to space is required, as we obviously recognize in Eq. (2.22). Special care has to be taken to solve this ill-conditioned problem in the presence of unavoidably noise (cf. Fig. 2.8.)Second, the estimation of the reaction fluxes and the diffusion coefficients in Eq. (2.22) by means of IMI has to be done simultaneous. Finally, the errors in the mass transport model will propagate in the estimation of the reaction flux expression. Hence, special care must be taken in the selection of the diffusion model structure. A final simultaneous identification step may also help in enhancing the confidence in the model parameters.

4.3.1 Validation in simulation and experiment

A model for the benzaldehydelyase (BAL) kinetics in the complete system was obtained (Zavrel et al., 2010). This was achieved by first investigating individual phenomena via experimental isolation and IMI. Finally, the complete model could be used to estimate all model parameters simultaneously. The comparison of the parameter estimates obtained for the individual and

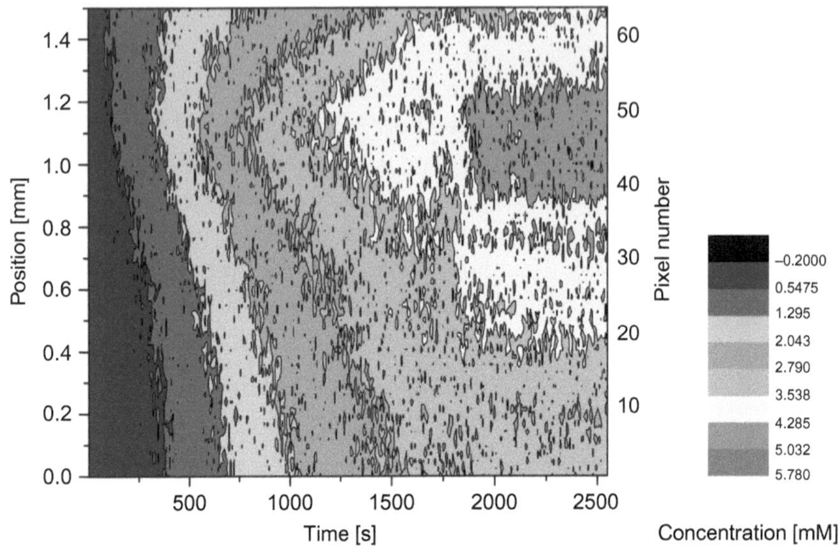

Figure 2.8 Temporal and spatial concentration gradients of DMBA in a κ-Carrageenan hydrogel bead. On the right axis the pixel number is shown, and on the left axis the corresponding position of the objective field of view in mm (Schwendt et al., 2010). *Copyright © (2010) Society for Applied Spectroscopy. Reprinted with permission. All rights reserved.*

coupled phenomena showed that kinetic phenomena may indeed interact. Hence, the common assumption that kinetic phenomena do not influence each other has been corrected.

5. IMI OF SYSTEMS WITH CONVECTIVE TRANSPORT

The applicability of IMI to relevant and challenging problems has been demonstrated in the previous sections. Still, the complexity tackled has been moderate, since three-dimensional (3D), transient transport and reaction problems in complex spatial geometries have not yet been treated. Such problems are relevant not only in chemical process systems, but in many other areas of science and engineering. As a first step toward the application of IMI to general 3D transient transport and reaction problems the identification of a transport coefficient function in the energy equation of a model of a wavy falling film (Karalashvili et al., 2008, 2011) and of a heat flux distribution during pool boiling (Heng, 2011; Lüttich et al., 2006) have been investigated.

5.1. Modeling of energy transport in falling liquid films

Falling liquid films are widely used in chemical engineering, for example, to implement coolers, evaporators, absorbers, or chemical reactors, where the wavy surface patterns are exploited to intensify heat and mass transfer between the liquid film and the surrounding gas. Even the dynamics of heated falling films of a single chemical species is complex and has been the subject of intensive research (e.g., Meza and Balakotaiah, 2008; Trevelyan et al., 2007). Direct numerical simulation of the free-surface, mixed initial-boundary problem involving the continuity, the momentum and the energy equations is very involved and has not yet been reported to the author's knowledge. Even if it were possible, the computational load would prevent its application for the design of technical equipment. As an alternative, Wilke (1962) suggested a long time ago to approximate the complex spatial domain of the wavy liquid film by a flat-film geometry and to introduce a so-called *effective transport coefficient* which has to account for the wave-induced back mixing present in the wavy film (Adomeit and Renz, 2000). Yet, there are no accepted and reasonably general models available which correlate the effective transport coefficient with the velocity and temperature fields in the falling film.

The IMI procedure seems to be a promising starting point to tackle this long-standing problem by the sequence of steps outlined in Section 3.1. The following exposition is based on the work of Karalashvili et al. (2008, 2011) and Karalashvili (2012).

5.1.1 Diffusive energy flux estimation (IMI.1–IMI.3)

The energy transport in a 3D, transient, flat falling film (cf. Fig. 2.9) can be represented by the energy equation, which can be reformulated for incompressible fluids (with constant density ρ) to result in

$$\rho\frac{\partial u(z,t)}{\partial t}=-\rho w(z,t)\cdot\nabla u(z,t)-\nabla\cdot j_u(z,t),\quad z\in\Omega,t>t_0 \qquad [2.23]$$

with appropriate initial and boundary conditions. The velocity field $w(z,t)$ is assumed to be known (either measured or computed from a possibly approximate solution of the Navier–Stokes equations), while the internal energy $u(z,t)$ (or rather the temperature $T(z,t)$) is assumed to be measured at reasonable spatiotemporal resolution. This *model B* can be refined by decomposing the diffusive energy flux $j_u(z,t)$ into a known molecular and an *unknown wave-induced* term. This reformulation results finally in

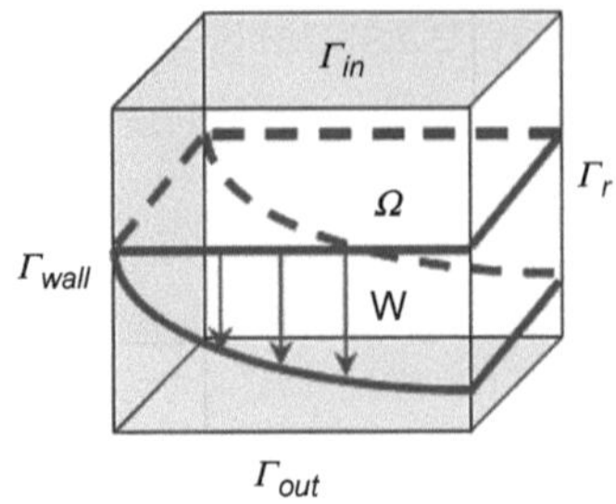

Figure 2.9 The geometry of the flat-film. *Copyright © (2011) Society for Industrial and Applied Mathematics. Reprinted with permission. All rights reserved.* (For color version of this figure, the reader is referred to the online version of this chapter.)

$$\frac{\partial T}{\partial t}+w\cdot\nabla T-\nabla\cdot(a_{mol}\nabla T)=f_w, \qquad [2.24]$$

with the known molecular transport coefficient a_{mol} and the unknown wavy contribution to the energy flux $f_w(z,t)$. This flux contribution can be reconstructed from temperature field data by solving a source inverse problem which is linear in the unkown $f_w(z,t)$ by an appropriate regularized numerical method (Karalashvili et al., 2008). Using (optimal) experiment design techniques, appropriate initial and boundary conditions may be found, which maximize the model identifiability.

5.1.2 Wavy energy flux model (IMI.4)

A reasonable model for the wavy contribution to the energy flux is motivated by Fourier's law. Hence, the flux $f_w(z,t)$ in Eq. (2.24) can be related to a wavy transport coefficient $a_w(z,t)$ by the Ansatz

$$f_w=-\nabla\cdot(a_w\nabla T),\quad z\in\Omega, t>t_0 \qquad [2.25]$$

Note, that the sum of the molecular and the wavy transport coefficients define an effective transport coefficient, that is, $a_{eff}=a_{mol}+a_w$. In order to estimate $a_w(z,t)$, a (nonlinear) coefficient inverse problem in the spatial domain has to be solved for any point in time t (Karalashvili et al., 2008).

5.1.3 Reducing the bias (IMI.5)

The *model BF* is formed by introducing Eq. (2.25) into Eq. (2.24). The resulting equation is used to reestimate the wavy coefficient $a_w(z,t)$ starting from the estimate in step *IMI.4* as initial values (Karalashvili et al., 2011).

5.1.4 Models for the wavy energy transport coefficient (IMI.6 and IMI.7)

A set of algebraic models is introduced to parameterize the transport coefficients in time and space by an appropriate model structure given as

$$a_w = m_{w,l}(z, t, \theta_l), \quad l \in \Sigma. \tag{2.26}$$

This set is the starting point for the identification of a suitable parametric model which properly relates the transport coefficient with velocity and temperature and possibly their gradients. The bias can again be removed by first inserting Eq. (2.26) into Eq. (2.25), and the result into Eq. (2.24) in order to reestimate the parameters prior to a ranking of the models with respect to model quality (Karalashvili et al., 2011). To measure the model quality and to select a "best-performing" transport model in a set of candidates Σ, we use AIC (Akaike, 1973). The model with minimum AIC is selected. Consequently, this criterion chooses models with the best fit of the data, and hence high precision in the parameters, but at the same time penalizes the number of model parameters.

5.1.5 Selecting the best transport coefficient model (IMI.8 and IMI.9)

An optimal design of experiments should finally be employed to obtain most informative measurements to finally identify the best model for $a_w(z,t)$ (Karalashvili and Marquardt, 2010).

5.1.6 Validation in simulation

We consider an illustrative flat-film case study without incorporating a priori knowledge on the unknown transport (Karalashvili et al., 2011). A convection–diffusion system describing energy transport in a single component fluid of density ρ on a flat domain $\Omega = (0, 1)^3[\text{mm}^3]$ is investigated. The boundary Γ consists of the inflow $\Gamma_{in} = \{z_1 = 0\}$, the outflow $\Gamma_{out} = \{z_1 = 1\}$, the wall $\Gamma_{wall} = \{z_2 = 0\}$ as well as the remaining boundaries Γ_r (cf. Fig. 2.9). Here, the spatial coordinate z_1 corresponds to the flow direction of the falling film, z_2 is the direction in the film thickness, and z_3 is the direction along the film width.

The density ρ and the heat capacity $c$are assumed to be constants. The velocity is given by the 1D Nusselt profile, $w(z,t) = 4.2857(2z_2 - {z_2}^2)$. The initial condition is $T(z,0) = 15\,[°\text{C}], z \in \Omega$. Boundary conditions are

$$T_{in}(z,t) = -30 z_2 t + 15[°\text{C}], \quad (z,t) \in \Gamma_{in} \times [t_0, t_f] \text{ and}$$
$$T_{wall}(z,t) = \left[100\left(1 - \cos\left(\pi \frac{z_1}{2}\right)\right)\right] t + 15[°\text{C}], \quad (z,t) \in \Gamma_{wall} \times [t_0, t_f].$$

At the other boundaries Γ_{out} and Γ_r, a zero flux condition is used. In this simulation experiment, the effective transport coefficient a_{eff} comprises a constant molecular term $a_{mol}=0.35\,[\text{mm}^2\text{s}]$ and a wavy transport term

$$\bar{a}_w=5\left(\bar{\vartheta}_1+\bar{\vartheta}_2 z_2 \sin\left(\bar{\vartheta}_3 z_1+\bar{\vartheta}_4 t\right)+\bar{\vartheta}_5 z_1 z_2+\bar{\vartheta}_6 z_1 z_2 z_3\right),$$
$$(z,t)\in\Omega\times\left[t_0,t_f\right] \qquad [2.27]$$

with the exact parameter values

$$\bar{\theta}=\left(\bar{\vartheta}_1,\bar{\vartheta}_2,\bar{\vartheta}_3,\bar{\vartheta}_4,\bar{\vartheta}_5,\bar{\vartheta}_6\right)^T=(1.1,1,\pi,0.02,0.2,0.02)^T. \qquad [2.28]$$

Motivated by physical considerations, a sinusoidal pattern has been chosen in the flow direction of the falling film. The time dependency is introduced such that the waves travel along the flow direction z_1 and propagate along the other directions, with a larger gradient in the z_2-direction (film thickness) and a relatively small gradient in the z_3-direction (film width).

High-quality temperature simulation data are generated by solving the linear problems (2.24), (2.25) with the exact transport model (2.27), (2.28) on a uniform fine grid with the spatial discretization consisting of $48\times48\times38$ intervals in the z_1-, z_2-, and z_3- directions, respectively. This yields a space discretization with 89,856 unknowns and 525,312 tetrahedral. As measurement data, we use the temperature data on a coarser grid with $24\times24\times19$ intervals to avoid the so-called inverse crime. For the time discretization, we use the implicit Euler scheme with time step $\tau=0.01$ s and apply 50 time steps starting from $t_0=0$ to $t_f=0.5\,[\text{s}]$. This results in 637,500 measurements. Furthermore, noisy measurements $\tilde{T}_m$ are generated by artificially perturbing the noise-free temperature T_m with measurement error ω, the values of which are generated from a zero mean normal distribution with variance one. Hence, we compute perturbed temperatures $\tilde{T}_m=\tilde{T}_m+\sigma\omega$ with standard deviation $\sigma=0.1$ of the measurement error.

Applying IMI, we compute an estimate $\hat{a}_w(z,t)$ of the wavy thermal diffusivity by solving the inverse problems (2.24) and (2.25), which has to be appropriately regularized to prevent undesirable amplification of measurement noise. These problems are formulated as optimization tasks and solved using adapted numerical iterative methods with appropriate stopping rules (cf. Karalashvili et al., 2011, for details). Figure 2.10 shows the wavy thermal diffusivity resulting from the second step *BF* at time instance $t=0.01$ s and constant $z_3=0.5$ mm. As can be seen, the chosen constant initial guess is very different from the true solution. Since the reconstruction of the wavy

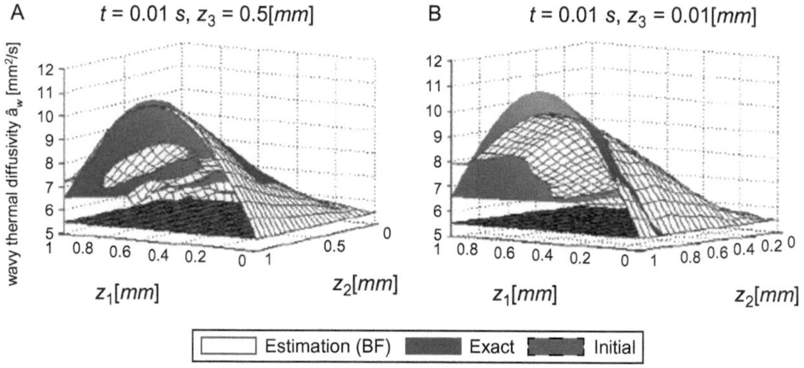

Figure 2.10 True and estimated wavy thermal diffusivity. *Copyright © (2011) Society for Industrial and Applied Mathematics. Reprinted with permission. All rights reserved.*

transport coefficient in Eq. (2.25) is decoupled in time, the obtained optimal solution at time instance $t=0.01$ s serves as a good initial value for the efficient optimization at later times. By exploring the shape of the reconstructed wavy transport coefficient $\hat{a}_w(z,t)$ (cf. Fig. 2.10), we develop a list Σ of model structures $m_{w,l}(z,t,\theta_l)$, $l \in \Sigma$. The estimate $\hat{a}_w(z,t)$ suggests that a reasonable model structure should incorporate a trigonometric function in the flow direction with a periodic change in time. Based on these observations, we propose a set of six candidate models as listed in Table 2.2. Obviously, the choice of model candidates requires intuition and physical insight. However, this choice can be efficiently guided by the results of the transport coefficient estimation step of the incremental identification method.

In the third step of the IMI, the parameters for each candidate model in Eq. (2.26) are estimated by using $\hat{a}_w(z,t)$ as inferential measurement data. The AIC values of the candidate models resulting from a multistart strategy are listed in the last column of Table 2.2 for noise-free and noisy measurements. In the presence of noise, the AIC values are significantly larger for all candidate models. The candidate models 4, 5, and 6 which employ an incorrect model structure, are of poor quality. Hence, the subset $\Sigma_s=\{1,2,3\}$ of reasonable model structures is left. The model of best quality obtained directly from IMI is candidate 1 (cf. AIC values in Table 2.2), which is the correct model. The corresponding optimal parameter vector is

$$\hat{\theta}_1=(1.140,0.803,4.077,-0.112,0.989,0.0336)^T. \qquad [2.29]$$

A comparison with the exact parameter vector (Eq. 2.28) shows that the deviation in the parameters is not the same for all parameters. Moreover, it is more

Table 2.2 Candidate models for all reactions wavy energy transport coefficient with corresponding values of the AIC

l	$m_{w,l}(z,t,\theta_l)$, $l \in \Sigma \cdot \{1,\ldots,6\}$	AIC/10^6 noise − free	AIC/10^6 noisy
1	$m_{w,1}=5(\vartheta_1+\vartheta_2 z_2 \sin(\vartheta_3 z_1+\vartheta_4 t)+\vartheta_5 z_1 z_2+\vartheta_6 z_1 z_2 z_3)$	−0.194	0.4272
2	$m_{w,2}=5(\vartheta_1+\vartheta_2 z_2 \sin(\vartheta_3 z_1+\vartheta_4 t))$	−0.112	0.6467
3	$m_{w,3}=5(\vartheta_1+\vartheta_2 z_2 \sin(\vartheta_3 z_1+\vartheta_4 t)+\vartheta_5 z_1 z_2)$	−0.184	0.4289
4	$m_{w,4}=5(\vartheta_1+\vartheta_3 z_1^2+\vartheta_4 t+\vartheta_5 z_1 z_2)$	1.785	1.9362
5	$m_{w,5}=5(\vartheta_1+\sin(\vartheta_3 z_1+\vartheta_4 t))$	2.210	2.2432
6	$m_{w,6}=5(\vartheta_1+\cos(\vartheta_3 z_1+\vartheta_4 t))$	2.334	2.3892

significant than the result obtained using noise-free data (Karalashvili et al., 2011). The reason for this is the error in the wavy transport coefficient estimate $\hat{a}_w(z,t)$, which is significantly larger compared to the one obtained from noise-free data (cf. Fig. 2.10A). However, despite the measurement noise, the same model structure as in the noise-free case can be recovered. This result shows, in fact, how difficult the solution of such ill-posed identification problems is if (inevitable) noise is present in the measurements. Though in the considered case the choice of the best model structure is not sensitive to noise, the quality of the estimated parameters deteriorates significantly despite the favorable situation that the correct model structure was in the set of candidates.

In order to reduce the inherent bias, we estimate in the correction procedure the parameters of each reasonable candidate model in subset $\Sigma_s = \{1,2,3\}$. Besides the corresponding optimal values of parameters available from the IMI procedure, an additional 500 randomly chosen initial values are used. The resulting AIC values for each of these candidates at their corrected optima indicate that candidate 1 is the "best performing" one. Figure 2.11 depicts the estimation result in comparison to the exact transport coefficient. The corresponding corrected optimal parameter vector results now in

$$\hat{\theta}_1 = (1.104, 0.723, 4.069, -0.149, 0.826, 0.186)^T. \qquad [2.30]$$

A comparison with the parameter estimates (Eq. 2.29) that follow directly after the IMI reveals that most of the parameter estimates are moved toward the exact parameter values (Eq. 2.28). Note that the fourth parameter

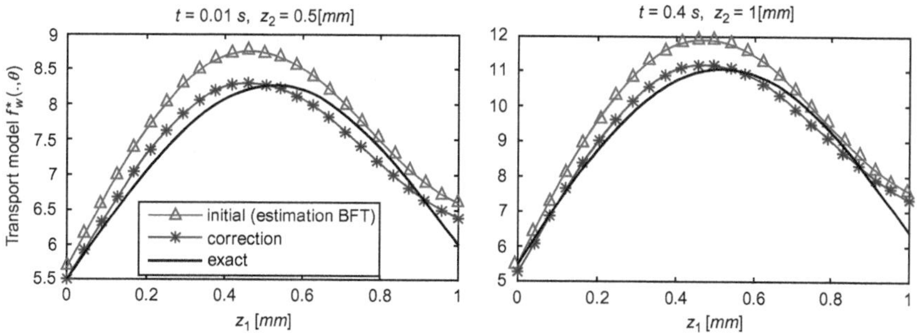

Figure 2.11 Estimation result in comparison to the exact and initial transport coefficient. *Copyright © (2011) Society for Industrial and Applied Mathematics. Reprinted with permission. All rights reserved.* (For color version of this figure, the reader is referred to the online version of this chapter.)

showing large deviations from the correct value governs the time dependency in the model structure. Because of the short duration of the experiment and the measurement noise, it cannot be correctly recovered.

An attempt to use the SMI approach for the direct parameter estimation problem with balance equation and model structure candidate 1 failed to converge. Convergence could not be achieved using the same initial values employed in the third step of the IMI method. Consequently, the IMI approach represents an attractive strategy to handle nonlinear, ill-posed, transient, distributed (3D) parameter systems with structural model uncertainty.

5.1.7 Experimental validation

It has not yet been accomplished. For one, the development of this variant of IMI has not yet been completed. Furthermore, high-resolution measurements of film thickness, temperature, and velocity fields are mandatory. Optical techniques are under investigation in collaborating research groups (Schagen et al., 2006). Moreover, the IMI is being investigated for the identification of effective mass transport models in falling film flows (Bandi et al., 2011). The model identification is based on high-resolution concentration measurements of oxygen being physically absorbed into an aqueous film. A planar laser-induced luminescence measurement technique is applied. It enables to simultaneously measure the 2D concentration distribution and the film thickness. The unique feature of this joint research work is the strong interaction between modeling, measurement techniques, and numerical simulation.

5.2. Heat flux estimation in pool boiling

In the study of boiling heat transfer, most research has been devoted in the past to the experimental investigation of the boiling heat flux averaged over the observation time and the heater surface. Numerous papers have contributed to the modeling of boiling heat transfer on the macro-scale, the meso- and microscopic as well as the molecular scale (Dhir and Liaw, 1989; Stephan and Hammer, 1994). There are also many works related to the numerical simulation of boiling processes, for example, Dhir (2001). Despite a lot of progress in understanding the physical fundamentals of boiling, current design methods are still mostly based on correlations which are valid only for one particular boiling regime. The parameters dominating the boiling heat transfer are unclear yet. Only by clarifying the mechanisms of heat transfer, vapor generation and two-phase flow phenomena such as the interfacial dynamics and the wetting structure as well as their interaction very close and at the boiling surface, substantial progress in the understanding of boiling processes can be accomplished.

The goal of our work in this area is to develop physically sound models of pool boiling processes and to identify major physical effects on various degrees of detail based on well-designed experiments. These models should at least achieve a qualitative and—as far as possible—a quantitative mechanistic prediction of the boiling heat flux as a function of the relevant system parameters. The overall system consisting of the two-phase vapor–liquid layer, the boiling surface and the heated wall close to the surface is schematically shown in Fig. 2.12. For an accurate modeling and analysis of boiling heat transfer mechanisms over the entire range of boiling conditions, the observation of local heat flux distribution on the boiling surface or its reconstruction from indirect measurements is an indispensable prerequisite. In combination with other measurements (like optical probes), which can be used to identify the interfacial geometry of the two-phase flow, the estimated transient local boiling heat flux distribution can be used for the development of physically sound heat transfer models for boiling regimes beyond low heat flux nucleate boiling, where heat transfer models can be derived from the study of single undisturbed bubbles.

Only a combined investigation of the mechanisms in the involved subsystems will allow the identification and meaningful interpretation of the relevant heat transfer phenomena. It currently seems impossible to infer the heat transfer characteristic of the whole boiling process from very detailed models simply because of computational complexity.

Hence, we first approach the estimation of the state at the boiling surface from the measurements inside the heater or the accessible surface in the sense of the IMI procedure. We consider the heat conduction inside the domain Ω (the test heater) which obeys the linear heat equation without sources with appropriate boundary and initial conditions, that is, Eq. (2.1) reduces to

$$\begin{aligned} \frac{\partial T(z,t)}{\partial t} &= -\nabla_z \cdot (\alpha \nabla_z T), \quad z \in \Omega, t > t_0, \\ T(z,t_0) &= T_0(z), \\ \nabla_z T|_{\partial\Omega} &= j_{b,y}, \quad z \in \partial\Omega. \end{aligned} \qquad [2.31]$$

The coefficient α denotes the thermal diffusivity and $T(z,t)$ the temperature field inside the heater. Since the variation of the temperature T throughout Ω is only within a few Kelvin, it suffices to assume that α is not dependent on the temperature. However, they may be functions of spatial coordinates, since Ω constitutes of some layers of different materials. In the actual experiments at TU Berlin (Buchholz et al., 2004) and TU Darmstadt (Wagner et al., 2007), distinct local temperature fluctuations are measured immediately below the surface by an array of microthermocouples or using an IR-camera. The measured temperature fluctuations inside the heater are an obvious consequence of the local heat flux $j_{b,y}$ and temperature fluctuations resulting from the wetting dynamics at the surface boundary of the heater which cannot be measured directly in order not to disturb the boiling process.

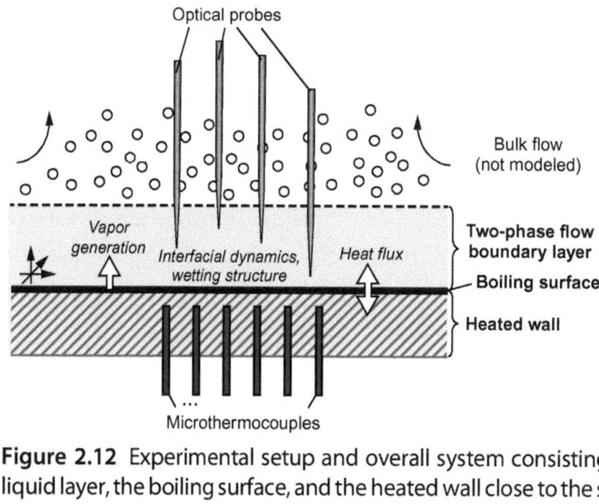

Figure 2.12 Experimental setup and overall system consisting of the two-phase vapor–liquid layer, the boiling surface, and the heated wall close to the surface (Lüttich et al., 2006).

Following the IMI procedure, the surface heat flux fluctuations $j_{b,y}$ could be identified from the measured temperature data in the different boiling regimes in the first step. The estimated surface heat flux and temperature may then serve in the next steps to identify a (physically motivated) correlation between them.

The heat flux estimation task, that is, the identification of the surface heat fluxes, is formulated as a 3D inverse heat conduction problem (IHCP) in the form of a regularized least-squares optimization. The resulting large-scale ill-posed problems were considered as computationally intractable for a long time (Lüttich et al., 2006). Although, there have been many attempts in the past to solve these kinds of IHCP, none of the available algorithms has been able to solve realistic problems (thick heaters, 3D, complex geometry, composite materials, real temperature sensor configurations, etc.) relevant to boiling heat transfer with high estimation quality.

Fortunately, our research group has been able to develop efficient and stable numerical solution techniques in recent years. In particular, Heng et al. (2008) have reconstructed local heat fluxes at three operating points along the boiling curve of isopropanol for the first time by using a simplified 3D geometry model and an optimization-based solution approach. The total computation took a few days on a normal PC. This approach was also applied to the reconstruction of local boiling heat flux in a single-bubble nucleate boiling experiment from a high-resolution temperature field measured at the back side of a thin heating foil (Heng et al., 2010). An efficient CGNE-based iterative regularization strategy has been presented by Egger et al. (2009) to particularly resolve the nonuniqueness of the solution resulting from limited temperature observations obtained in the experiment of Buchholz et al. (2004). Moreover, a space-time finite-element method was used to allow a fast numerical solution of the arising direct, adjoint, and sensitivity problems, which for the first time facilitated the treatment of the entire heater in 3D. The computational efficiency could be improved, such that an estimation task of similar size required only several hours of computational time. However, this kind of approach is still restricted to a fixed uniform discretization. Since the boiling heat flux is nonuniformly distributed on the heater surface due to the strong local activity of the boiling process, an adaptive mesh refinement strategy is an appropriate choice for further method improvement. As a first step toward a fully adaptive spatial discretization of the inverse boiling problem, multilevel adaptive methods via temperature-based as well as heat flux-based error estimation techniques have been developed recently (Heng et al., 2010). The proposed multilevel adaptive iterative regularization method can treat both spatially highly resolved and point-wise temperature measurements very efficiently, independent of the chosen boiling fluid and the shape of the heater.

5.2.1 Validation in simulation and experiment

The estimation and investigation of local boiling heat flux distribution by means of 3D heater geometry models has been performed for two different real pool boiling experiments. While one experiment (Wagner et al., 2007) has been conducted to generate single-bubble boiling processes which is technically only reasonable for low and intermediate heat fluxes, the other experiment (Buchholz et al., 2004) has been conducted on a technically relevant thick heater, which has been designed to observe the local phenomena for all boiling regimes. Figure 2.13 shows, for example, the estimation results obtained for a single-bubble experiment. From these results, it is apparent that the boiling heat flux undergoes a significant change during this single-bubble cycle and an interesting ring-shaped local heat flux is observed. The peak value of the estimated heat fluxes appears in the ring region and is nearly 30 times larger than the average value. We obtained similar results for other fluids and fluid mixtures. These estimation results represent a step toward the confirmation of the microlayer theory (Stephan and Hammer, 1994), which predicts that most of the heat during boiling is transferred in the microregion of the three-phase contact line by evaporation.

6. INCREMENTAL VERSUS SIMULTANEOUS IDENTIFICATION

In contrast to SMI, the IMI approach explicitly accounts for the fact that often an appropriate structure of one or more submodels in a complex process systems model is uncertain. The selection of the most suitable submodel structure has to be considered an integral part of the model identification process. Since model identification cannot be reduced to estimating the parameters from most informative experiments in a given, identifiable model structure, the model (structure) identification process has to be fully transparent to the modeler. Partial prior knowledge regarding model structure can easily be incorporated. Missing submodels are derived either from experimental or from inferred input–output data in the previous estimation step supported by theoretical investigations on a finer (often the molecular) scale. Any decision on the model structure relates to a single physicochemical phenomenon and thus reduces ambiguity. Identifiability can be assessed

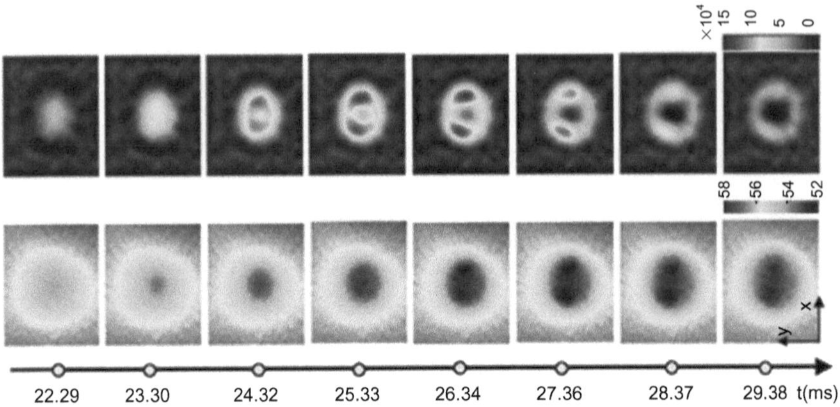

Figure 2.13 The measured temperature field on the back side of the thin heating foil and the estimated surface boiling heat flux at given times (Heng et al., 2010). *Copyright © (2010) Taylor & Francis. Reprinted with permission. All rights reserved.* (For color version of this figure, the reader is referred to the online version of this chapter.)

more easily on the level of the submodel. This way, the IMI strategy supports the discovery of novel model structures which are consistent with the available experimental data.

The decomposition strategy of IMI is also very favorable from a computational perspective. It drastically reduces computational load, because it breaks the curse of dimensionality due to the combinatorial nature of the decision making problem related to submodel selection. IMI avoids this problem, because the decision making is integrated into the decomposition strategy and systematically exploits knowledge acquired during the previous identification steps. Furthermore, the computational effort is reduced because the solution of a *strongly nonlinear inverse problem* involving (partial) differential–algebraic equations is replaced by a sequence of less complex, often *linear inverse problems* and a few *algebraic regression problems.* This divide-and-conquer approach also improves the robustness of the numerical algorithms and their sensitivity toward the choice of initial estimates. Last but not least, the decomposition strategy facilitates quasi-global parameter estimation in those cases where all but the last nonlinear regression problem are convex. A general quasi-global deterministic solution strategy is worked out by Michalik et al. (2009a,b,c,d) for identification problems involving differential–algebraic problems.

The computational advantages of IMI become decisive in case of the identification of complex 3D transport and reaction models on complex spatial domains. Our case studies indicate, that SMI is computationally often intractable while IMI renders the estimation problems feasible or at least reduces the load by orders of magnitude. Identifiability analysis and optimal design of experiments are key to success in case of 3D transport and reaction problems, because sufficient excitation in time and space can typically not be achieved intuitively.

Error propagation is unavoidable in IMI, because any estimation error will impair the estimation quality in the following steps. The resulting bias can, however, be easily removed by a final correction step, where a parameter estimation problem is solved for the best aggregated model(s) using very good initial parameter values. Convergence is typically achieved in one or very few iterations.

Both, IMI and SMI are not successful, if the information content of the measurements is insufficient. However, identifiability problems can be discovered and remedied more easily in IMI compared to SMI. Then, either the model has to be simplified (to result in less unknown model parameters) or additional sensors have to be installed in the experiment.

7. CONCLUDING DISCUSSION

The exemplary applications of IMI as an integral part of the MEXA work process section not only demonstrate its versatility but also its distinct advantages compared to established SMI methods (Bardow and Marquardt, 2004a,b).

Our experience in a wide area of applications shows that a sensible integration of modeling and experimentation is indispensible if the mathematical model is supposed to extrapolate with adequate accuracy well beyond the region where model identification has been carried out. Such good extrapolation provides at least an indication that the physicochemical mechanisms underlying the observed system behavior have been captured by the model to a certain extent.

A coordinated design of the model structure and the experiment as advocated in the MEXA work process is most appropriate for several reasons (cf. Bard, 1974; Beck and Woodbury, 1998; Iyengar and Rao, 1983; Kittrell, 1990). On the one hand, an overly detailed model is often not identifiable even if perfect measurements of all the state variables were available (cf. Quaiser and Mönnigmann (2009) for an example from systems biology). Hence, any model should only cover a level of detail, which facilitates an experimental investigation of model validity. On the other hand, an overly simplified model does often not reflect real behavior satisfactorily. For example, equilibrium tray models in distillation assume phase equilibrium rather than accounting for the mass transfer resistance between the liquid and vapor phases. Though this model is still widely used in industrial practice, it has been shown to be inconsistent with basic physical principles, since it does not reflect the cross-effects of multicomponent diffusion (Taylor and Krishna, 1993). Such a coordinated design of experiment and models is closely related to the requirement of refining a model only based on experimental evidence (Markus et al., 1981). In particular, if a model is able to predict the accessible observations on the associated real system sufficiently well, its further refinement cannot be justified because it reduces the level of confidence in the model.

The identification of useful models at minimal effort requires a multidisciplinary team effort. Experts in high-resolution measurement techniques, the application domain of interest, numerical analysis, and modeling methodologies have to join forces to leverage the very high effort of model identification. Best-practices and suitable software environments, tailored to a certain application, such as reaction kinetics identification seem to be indispensable to roll out the MEXA framework into routine application.

ACKNOWLEDGMENTS

This work has been carried out as part of CRC 540 "Model-based Experimental Analysis of Fluid Multi-Phase Reactive Systems" which has been funded by the German Research Foundation (DFG) from 1999 to 2009. The substantial financial support of DFG is gratefully acknowledged. Furthermore, the contributions of the CRC 540 team, in particular, however of A. Bardow, M. Brendel, M. Karalashvili, E. Kriesten, C. Michalik, Y. Heng, and N. Kerimoglu are appreciated.

REFERENCES

Adomeit P, Renz U: Hydrodynamics of three-dimensional waves in laminar falling films, *Int J Multiphas Flow* 26(7):1183–1208, 2000.

Agarwal M: Combining neural and conventional paradigms for modelling, prediction and control, *Int J Syst Sci* 28:65–81, 1997.

Akaike H: Information theory as an extension of the maximum likelihood principle. In Petrov BN, Csaki F, editors: *Second international symposium on information theory*, Budapest, 1973, Akademiai Kiado, pp 267–281.

Alsmeyer F, Koß H-J, Marquardt W: Indirect spectral hard modeling for the analysis of reactive and interacting mixtures, *J Appl Spectrosc* 58(8):975–985, 2004.

Amrhein M, Bhatt N, Srinivasan B, Bonvin D: Extents of reaction and flow for homogeneous reaction systems with inlet and outlet streams, *AIChE J* 56(11):2873–2886, 2010.

Ansorge-Schumacher M, Greiner L, Schroeper F, Mirtschin S, Hischer T: Operational concept for the improved synthesis of (R)-3,3'-furoin andrelated hydrophobic compounds with benzaldehydelyase, *Biotechnol J* 1(5):564–568, 2006.

Asprey SP, Macchietto S: Statistical tools in optimal model building, *Comput Chem Eng* 24:1261–1267, 2000.

Balsa-Canto E, Banga JR: AMIGO: a model identification toolbox based on global optimization and its applications in biosystems. In *11th IFAC symposium on computer applications in biotechnology, Leuven, Belgium*, 2010.

Bandi P, Pirnay H, Zhang L, et al: Experimental identification of effective mass transport models in falling film flows. In *6th International Berlin workshop (IBW6) on transport phenomena with moving boundaries, Berlin*, 2011.

Bard Y: *Nonlinear parameter estimation,* 1974, Academic Press.

Bardow A: *Model-based experimental analysis of multicomponent diffusion in liquids,* Düsseldorf, 2004, VDI-Verlag (Fortschritt-Berichte VDI: Reihe 3, Nr. 821).

Bardow A, Marquardt W: Identification of diffusive transport by means of an incremental approach, *Comput Chem Eng* 28(5):585–595, 2004a.

Bardow A, Marquardt W: Incremental and simultaneous identification of reaction kinetics: methods and comparison, *Chem Eng Sci* 59(13):2673–2684, 2004b.

Bardow A, Marquardt W: Identification methods for reaction kinetics and transport. In Floudas CA, Pardalos PM, editors: *Encyclopedia of optimization,* ed 2, 2009, Springer, pp 1549–1556.

Bardow A, Marquardt W, Göke V, Koß HJ, Lucas K: Model-based measurement of diffusion using Raman spectroscopy, *AIChE J* 49(2):323–334, 2003.

Bardow A, Göke V, Koß H-J, Lucas K, Marquardt W: Concentration-dependent diffusion coefficients from a single experiment using model-based Raman spectroscopy, *Fluid Phase Equilib* 228–229:357–366, 2005.

Bardow A, Göke V, Koß HJ, Marquardt W: Ternary diffusivities by model-based analysis of Raman spectroscopy measurements, *AIChE J* 52(12):4004–4015, 2006.

Bardow A, Bischof C, Bücker M, et al: Sensitivity-based analysis of the k-e- model for the turbulent flow between two plates, *Chem Eng Sci* 63:4763–4776, 2008.

Bastin G, Dochain D: *On-line estimation and adaptive control of bioreactors,* Amsterdam, 1990, Elsevier.

Bauer M, Geyer R, Griengl H, Steiner W: The use of lewis cell to investigate the enzyme kinetics of an (s)-hydroxynitrilelyase in two-phase systems, *Food Technol Biotechnol* 40 (1):9–19, 2002.

Beck JV, Woodbury KA: Inverse problems and parameter estimation: integration of measurements and analysis, *Meas Sci Technol* 9(6):839–847, 1998.

Berendsen W, Lapin A, Reuss M: Investigations of reaction kinetics for immobilized enzymes—identification of parameters in the presence of diffusion limitation, *Biotechnol Prog* 22:1305–1312, 2006.

Berger RJ, Stitt E, Marin G, Kapteijn F, Moulijn J: Eurokin—chemical reaction kinetics in practice, *CatTech* 5(1):30–60, 2001.

Bhatt N, Amrhein M, Bonvin D: Extents of reaction, mass transfer and flow for gas-liquid reaction systems, *Ind Eng Chem Res* 49(17):7704–7717, 2010.

Bhatt N, Kerimoglu N, Amrhein M, Marquardt W, Bonvin D: Incremental model identification for reaction systems—a comparison of rate-based and extent-based approaches, *Chem Eng Sci* 83:24–38, 2012.

Biegler LT: *Nonlinear programming: concepts, algorithms, and applications to chemical processes,* Philadelphia, 2010, SIAM.

Bird RB: Five decades of transport phenomena, *AIChE J* 50(2):273–287, 2004.

Bird RB, Stewart WE, Lightfoot EN: *Transport phenomena,* ed 2, 2002, Wiley.

Bonvin D, Rippin DWT: Target factor analysis for the identification of stoichiometric models, *Chem Eng Sci* 45(12):3417–3426, 1990.

Bothe D, Lojewski A, Warnecke H-J: Computational analysis of an instantaneous irreversible reaction in a T-microreactor, *AIChE J* 56(6):1406–1415, 2010.

Brendel M, Marquardt W: Experimental design for the identification of hybrid reaction models from transient data, *Chem Eng J* 141:264–277, 2008.

Brendel M, Marquardt W: An algorithm for multivariate function estimation based on hierarchically refined sparse grids, *Comput Vis Sci* 12(4):137–153, 2009.

Brendel M, Bonvin D, Marquardt W: Incremental identification of kinetic models for homogeneous reaction systems, *Chem Eng Sci* 61:5404–5420, 2006.

Britt HI, Luecke RH: Parameter estimation with error in observables, *Am J Phys* 43(4):372, 1975.

Buchholz K: Immobilized enzymes—kinetics, efficiency, and applications, *Chem Ing Tech* 61 (8):611–620, 1989.

Buchholz M, Auracher H, Lüttich T, Marquardt W: Experimental investigation of local processes in pool boiling along the entire boiling curve, *Int J Heat Fluid Flow* 25 (2):243–261, 2004.

Burnham KP, Anderson DR: *Model selection and multimodel inference: a practical information-theoretic approach,* ed 2, New York, 2002, Springer.

Buzzi-Ferraris G, Manenti F: Kinetic models analysis, *Chem Eng Sci* 64(5):1061–1074, 2009.

Cannon JR, DuChateau P: An inverse problem for a nonlinear diffusion equation, *SIAM J Appl Math* 39:272–289, 1980.

Cheng ZM, Yuan WK: Initial estimation of heat transfer and kinetic parameters of a wall-cooled fixed-bed reactor, *Comput Chem Eng* 21(5):511–519, 1997.

Craven P, Wahba G: Smoothing noisy data with spline functions, *Numer Math* 31:377–403, 1979.

Dhir VK: Numerical simulations of pool-boiling heat transfer, *AIChE J* 47:813–834, 2001.

Dhir VK, Liaw SP: Framework for a unified model for nucleate and transition pool boiling, *J Heat Transf* 111:739–745, 1989.

Egger H, Heng Y, Marquardt W, Mhamdi A: Efficient solution of a three-dimensional inverse heat conduction problem in pool boiling, *Inverse Probl* 25(9):095006, 2009 (19 pp).

Engl HW, Hanke M, Neubauer A: *Regularization of inverse problems,* Dordrecht, 1996, Kluwer.

Engl HW, Flamm C, Kügler P, Lu J, Müller S, Schuster P: Inverse problems in systems biology, *Inverse Probl* 25:123014, 2009.

Franceschini G, Macchietto S: Model-based design of experiments for parameter precision: state of the art, *Chem Eng Sci* 63(19):4846–4872, 2008.

Froment GF, Bischoff KB: *Chemical reactor analysis and design,* New York, 1990, Wiley.

Golub GH, Heath M, Wahba G: Generalized cross validation as a method for choosing a good ridge parameter, *Technometrics* 21(2):215–223, 1979.

Halling P: Biocatalysis in multi-phase reaction mixtures containing organicliquids, *Biotechnol Adv* 5(1):47–84, 1987.

Hanke M: *Conjugate gradient type methods for Ill-posed problems,* Harlow, Essex, 1995, Longman.

Hanke M, Scherzer O: Error analysis of an equation method for the identification of the diffusion coefficient in a quasi-linear parabolic differential equation, *SIAM J Appl Math* 59:1012–1027, 1999.

Hansen PC: *Rank-defficient and discrete Ill-posed problems: NumericalAspects of linear inversion,* Philadelphia, 1998, SIAM.

Hansen PC: Regularization tools version 3.0 for matlab 5.2, *Numer Algorithms* 20 (3):195–196, 1999.

Hastie T, Tibshirani R, Friedman J: *The elements of statistical learning: data mining, inference, and prediction,* New York, 2003, Springer.

Heng Y: *Mathematical formulation and efficient solution of 3d inverse heat transfer problems in pool boiling,* Düsseldorf, 2011, VDI-Verlag (Fortschritt-Berichte VDI, Nr. 922).

Heng Y, Mhamdi A, Groß S, et al: Reconstruction of local heat fluxes in pool boiling experiments along the entire boiling curve from high resolution transient temperature measurements, *Int J Heat Mass Transf* 51(21–22):5072–5087, 2008.

Heng Y, Mhamdi A, Wagner E, Stephan P, Marquardt W: Estimation of local nucleate boiling heat flux using a three-dimensional transient heat conduction model, *Inverse Probl Sci Eng* 18(2):279–294, 2010.

Higham DJ: Modeling and simulating chemical reactions, *SIAM Rev* 50:347–368, 2008.

Hirschorn RM: Invertibility of nonlinear control systems, *SIAM J Control Optim* 17:289–297, 1979.

Hosten LH: A comparative study of short cut procedures for parameter estimation in differential equations, *Comput Chem Eng* 3:117–126, 1979.

Huang C: Boundary corrected cubic smoothing splines, *J Stat Comput Sim* 70:107–121, 2001.

Iyengar SS, Rao MS: Statistical techniques in modelling of complex systems—single and multiresponse models, *IEEE Trans Syst Man Cyb* 13(2):175–189, 1983.

Kahrs O, Marquardt W: Incremental identification of hybrid process models, *Comput Chem Eng* 32(4–5):694–705, 2008.

Kahrs O, Brendel M, Michalik C, Marquardt W: Incremental identification of hybrid models of process systems. In van den Hof PMJ, Scherer C, Heuberger PSC, editors: *Model-based control*, Dordrecht, 2009, Springer, pp 185–202.

Karalashvili M: *Incremental identification of transport phenomena in laminar wavy film flows*, Düsseldorf, 2012, VDI-Verlag (Fortschritt-Berichte VDI, Nr. 930).

Karalashvili M, Marquardt W: Incremental identification of transport models in falling films. In *International symposium on recent advances in chemical engineering, IIT Madras, December 2010*, 2010.

Karalashvili M, Groß S, Mhamdi A, Reusken A, Marquardt W: Incremental identification of transport coefficients in convection-diffusion systems, *SIAM J Sci Comput* 30 (6):3249–3269, 2008.

Karalashvili M, Groß S, Marquardt W, Mhamdi A, Reusken A: Identification of transport coefficient models in convection-diffusion equations, *SIAM J Sci Comput* 33 (1):303–327, 2011.

Kerimoglu N, Picard M, Mhamdi A, Grenier L, Leitner W, Marquardt W: Incremental model identification of reaction and mass transfer kinetics in a liquid-liquid reaction system—an experimental study. In *AICHE 2011, Minneapolis Convention Center Minneapolis, MN, USA*, 2011.

Kerimoglu N, Picard M, Mhamdi A, Greiner L, Leitner W, Marquardt W: Incremental identification of a full model of a Two-phase friedel-crafts acylation reaction. In *ISCRE 22, Maastricht, Netherlands*, 2012.

Kirsch A: *An introduction to the mathematical theorie of inverse problems*, New York, 1996, Springer.

Kittrell JR: Mathematical modelling of chemical reactions, *Adv Chem Eng* 8:97–183, 1970.

Klipp E, Herwig R, Kowald A, Wierling C, Lehrach H: *Systems biology in practice. Concepts, implementation, and application*, Weinheim, 2005, Wiley.

Körkel S, Kostina E, Bock HG, Schlöder JP: Numerical methods for optimal control problems in design of robust optimal experiments for nonlinear dynamic processes, *Optim Method Softw* 19(3–4):327–338, 2004.

Kriesten E, Alsmeyer F, Bardow A, Marquardt W: Fully automated indirect hard modeling of mixture spectra, *Chemometr Intell Lab Syst* 91:181–193, 2008.

Kriesten E, Voda MA, Bardow A, et al: Direct determination of the concentration dependence of diffusivities using combined model-based Raman and NMR experiments, *Fluid Phase Equilib* 277:96–106, 2009.

Lohmann T, Bock HG, Schlöder JP: Numerical methods for parameter estimation and optimal experiment design in chemical reaction systems, *Ind Eng Chem Res* 31(1):54–57, 1992.

Lüttich T, Marquardt W, Buchholz M, Auracher H: Identification ofunifying heat transfer mechanisms along the entire boiling curve, *Int J Therm Sci* 45(3):284–298, 2006.
Mahoney AW, Doyle FJ, Ramkrishna D: Inverse problems in population balances: growth and nucleation from dynamic data, *AIChE J* 48(5):981–990, 2002.
Markus M, Plesser T, Kohlmeier M: Analysis of progress curves in enzyme kinetics—bias and convergent set in the differential and in the integral method, *J Biochem Biophys Methods* 4(2):81–90, 1981.
Marquardt W: Towards a process modeling methodology. In Berber R, editor: *Methods of model-based control, NATO-ASI Ser. E, Applied Sciences,* 1995, Kluwer Press, pp S.3–S.41.
Marquardt W: Model-based experimental analysis of kinetic phenomena in multi-phase reactive systems, *Chem Eng Res Des* 83(A6):561–573, 2005.
Marquardt W: Identification of kinetic models by incremental refinement. In Gähde U, Hartmann S, Wolf JH, editors: *Models, simulations, and the reduction of complexity*, Berlin, 2013, Walter de Gruyter (in press).
Marquardt W, Wedel Lv, Bayer B: Perspectives on lifecycle process modeling, *AIChE Symp Ser* 323(96):192–214, 2000.
Mason RL, Gunst RF, Hess JL: *Statistical design and analysis of experiments—with applications to engineering and science,* ed 2, 2003, Wiley.
Meza CE, Balakotaiah V: Modeling and experimental studies of large amplitude waves on vertically falling films, *Chem Eng Sci* 63:4704–4734, 2008.
Mhamdi A, Marquardt W: An inversion approach for the estimation of reaction rates in chemical reactors. In *ECC'99, Karlsruhe*, 1999 (31.8.-3.9).
Michalik C, Schmidt T, Zavrel M, Ansorge-Schumacher M, Spieß A, Marquardt W: Application of the incremental identification method to the formate oxidation using formate dehydrogenase, *Chem Eng Sci* 62(3):5592–5597, 2007.
Michalik C, Stuckert M, Marquardt W: Optimal experimental design for discriminating numerous model candidates—the AWDC criterion, *Ind Eng Chem Res* 49(2):913–919, 2009a.
Michalik C, Chachuat B, Marquardt W: Incremental global parameter estimation in dynamical systems, *Ind Eng Chem Res* 48:5489–5497, 2009b.
Michalik C, Brendel M, Marquardt W: Incremental identification of fluid multi-phase reaction systems, *AIChE J* 55(4):1009–1022, 2009c.
Michalik C, Hannemann R, Marquardt W: Incremental single shooting—a robust method for the estimation of parameters in dynamical systems, *Comput Chem Eng* 33:1298–1305, 2009d.
Oliveira R: Combining first principles modelling and artificial neural networks: a general framework, *Comput Chem Eng* 28:755–766, 2004.
Pope SB: *Turbulent flows,* 2000, Cambridge Univ. Press.
Popper K: *The logic of scientific discovery,* London, 1959, Hutchinson.
Prausnitz JM, Lichtenthaler RN, Gomes de Azevedo E: *Molecular thermodynamics of fluid-phase equilibria,* ed 3, New Jersey, 2000, Prentice Hall.
Psichogios DC, Ungar LH: A hybrid neural network—first principles approach to process modeling, *AIChE J* 38:1499–1511, 1992.
Pukelsheim F: *Optimal design of experiments,* Philadelphia, 2006, SIAM.
Quaiser T, Mönnigmann M: Systematic identifiability testing for unambiguous mechanistic modeling—application to JAK-STAT, MAP kinase, and NF-kappa B signaling pathway models, *BMC Syst Biol* 3:50, 2009.
Quaiser T, Dittrich A, Schaper F, Mönnigmann M: A simple workflow for biologically inspired model reduction—application to early JAK-STAT signaling, *BMC Syst Biol* 5:30, 2011.
Ramsay JO: Functional components of variation in handwriting, *J Am Stat Assoc* 95 (449):9–15, 2000.

Ramsay JO, Ramsey JB: Functional data analysis of the dynamics of the monthly index of nondurable goods production, *J Econom* 107(1–2):327–344, 2002.
Ramsay JO, Munhall KG, Gracco VL, Ostry DJ: Functional data analyses of lip motion, *J Acoust Soc Am* 99(6):3718–3727, 1996.
Reinsch CH: Smoothing by spline functions, *Num Math* 10:177–183, 1967.
Romijn R, Özkan L, Weiland S, Ludlage J, Marquardt W: A grey-box modeling approach for the reduction of nonlinear systems, *J Process Control* 18(9):906–914, 2008.
Ruppen D: A contribution to the implementation of adaptive optimal operation for discontinuous chemical reactors. PhD thesis. ETH Zuerich, 1994.
Schagen A, Modigell M, Dietze G, Kneer R: Simultaneous measurement of local film thickness and temperature distribution in wavy liquid films using a luminescence technique, *Int J Heat Mass Transf* 49(25–26):5049–5061, 2006.
Schittkowski K: *Numerical data fitting in dynamical systems: a practical introduction with applications and software,* Dordrecht, 2002, Kluwer.
Schmidt T, Michalik C, Zavrel M, Spieß A, Marquardt W, Ansorge-Schumacher M: Mechanistic model for prediction of formate dehydrogenase kinetics under industrially relevant conditions, *Biotechnol Prog* 26:73–78, 2009.
Schwendt T, Michalik C, Zavrel M, et al: Determination of temporal and spatial concentration gradients in hydrogel beads using multiphoton microscopy techniques, *Appl Spectrosc* 64(7):720–726, 2010.
Slattery J: *Advanced transport phenomena,* Cambridge, 1999, Cambridge Univ. Press.
Stephan P, Hammer J: A new model for nucleate boiling heat transfer, *Wärme Stoffübertrag* 30 (2):119–125, 1994.
Stewart WE, Shon Y, Box GEP: Discrimination and goodness of fit of multiresponse mechanistic models, *AIChE J* 44:1404–1412, 1998.
Takamatsu: The nature and role of process systems engineering, *Comput Chem Eng* 7 (4):203–218, 1983.
Taylor R, Krishna R: *Multicomponent mass transfer,* New York, 1993, Wiley.
Telen D, Logist F, Van Derlinden E, Tack I, Van Impe J: Optimal experiment design for dynamic bioprocesses: a multi-objective approach, *Chem Eng Sci* 78:82–97, 2012.
Tholudur A, Ramirez WF: Neural-network modeling and optimization of induced foreign protein production, *AIChE J* 45(8):1660–1670, 1999.
Tikhonov AN, Arsenin VY: *Solution of Ill-posed problems*, Washington, 1977, V. H. Winston & Son.
Timmer J, Rust H, Horbelt W, Voss HU: Parametric, nonparametric and parametric modelling of a chaotic circuit time series, *Physics Lett A* 274(3–4):123–134, 2000.
Trevelyan PMJ, Scheid B, Ruyer-Quil C, Kalliadasis S: Heated falling films, *J Fluid Mech* 592:295–334, 2007.
Tyrell HJV, Harris KR: *Diffusion in liquids,* London, 1984, Butterworths.
Vajda S, Rabitz H, Walter E, Lecourtier Y: Qualitative and quantitative identifiability analysis of nonlinear chemical kinetic models, *Chem Eng Commun* 83:191–219, 1989.
Van Lith PF, Betlem BHL, Roffel B: A structured modelling approach for dynamic hybrid fuzzy-first principles models, *J Process Control* 12(5):605–615, 2002.
van Roon J, Arntz M, Kallenberg A, et al: A multicomponent reaction–diffusion model of a heterogeneously distributed immobilized enzyme, *Appl Microbiol Biotechnol* 72 (2):263–278, 2006.
Verheijen PJT: Model selection: an overview of practices in chemical engineering. In Asprey SP, Macchietto S, editors: *Dynamic model development: methods, theory and applications*, Amsterdam, 2003, Elsevier, pp 85–104.
Voss HU, Rust H, Horbelt W, Timmer J: A combined approach for the identification of continuous non-linear systems, *Int J Adapt Control Signal Process* 17(5):335–352, 2003.

Wagner E, Sprenger A, Stephan P, Koeppen O, Ziegler F, Auracher H: Nucleate boiling at single artificial cavities: bubble dynamics and local temperature measurements. In *Proceedings of 6th International Conference on Multiphase Flow. Leipzig, Germany*, 2007.

Wahl SA, Haunschild MD, Oldiges M, Wiechert W: Unravelling the regulatory structure of biochemical networks using stimulus response experiments and large-scale model selection, *IEE Proc Syst Biol* 153(4):275–285, 2006.

Walter E, Pronzato L: Qualitative and quantitative experiment design for phenomenological models—a survey, *Automatica* 26(2):195–213, 1990.

Walter E, Pronzato L: *Identification of parametric models from experimental data,* Berlin, 1997, Springer.

Wilke W: *Wärmeübergang an Rieselfilmen,* Düsseldorf, 1962, VDI-Verlag (VDI-Forsch.-Heft 490).

Zavrel M, Michalik C, Schwendt T, et al: Systematic determination of intrinsic reaction parameters in enzyme immobilizates, *Chem Eng Sci* 65(8):2491–2499, 2010.

CHAPTER THREE

Wavelets Applications in Modeling and Control

Arun K. Tangirala*, Siddhartha Mukhopadhyay†, Akhilanand P. Tiwari‡
*Department of Chemical Engineering, IIT Madras, Chennai, Tamil Nadu, India
†Bhabha Atomic Research Centre, Control Instrumentation Division, Mumbai, India
‡Bhabha Atomic Research Centre, Reactor Control Division, Mumbai, India

Contents

Advances in Chemical Engineering, Volume 43
ISSN 0065-2377
http://dx.doi.org/10.1016/B978-0-12-396524-0.00003-9

Abstract

Wavelets have been on the forefront for more than three decades now. Wavelet transforms have had tremendous impact on the fields of signal processing, signal coding, estimation, pattern recognition, applied sciences, process systems engineering, econometrics, and medicine. Built on these transforms are powerful frameworks and novel techniques for solving a large class of theoretical and industrial problems. Wavelet transforms facilitate a *multiscale* framework for signal and system analysis. In a multiscale framework, the analyst can decompose signals into components at different resolutions followed by the application of the standard single-scale techniques to each of these components. In the area of process systems engineering, wavelets have become the de facto tool for signal compression, estimation, filtering, and identification. The field of wavelets is ever-growing with invaluable and innovative contributions from researchers worldwide. The purpose of this chapter is threefold: (i) to provide a semi-formal introduction to wavelet transforms for engineers; (ii) to present an overview of their applications in process systems engineering, with specific attention to controller loop performance monitoring and empirical modeling; and (iii) to introduce the ideas of consistent prediction-based multiscale identification. Case studies and examples are used to demonstrate the concepts and developments in this work.

1. INTRODUCTION

1.1. Motivation

Every process that we come across, natural or man-made, is characterized by a mixture of phenomena that evolve at different timescales. The term *timescale* often refers to the pace or rate at which the associated subsystem changes whenever the system is subjected to an internal or an external perturbation. Due to the differences in their rates of evolution, certain subsystems settle faster or slower than the remaining. Needless to say, the slowest subsystem governs the settling time of the overall system. Systems with such characteristics are known as *multiscale* systems. In contrast, a *single-scale* system operates at a single evolution rate. Multiscale systems are ubiquitous—they are encountered in all spheres of sciences and engineering (Ricardez-Sandoval, 2011; Vlachos, 2005). In chemical engineering, the two time-constant (time-scale) process is a classical example of a multiscale system (Christofides and Daoutidis, 1996). Measurements of process variables contain contributions from subsystems and (instrumentation) devices with significantly different time constants. A fuel cell system (Frano, 2005) exhibits multiscale behavior due to the large differences in the time-scales of the electrochemical subsystem (order of 10^{-5} s), the fuel flow subsystem (order of 10^{-1} s), and the thermal subsystem (order of 10^2–10^3 s). The atmospheric system is a complex, large, multiscale system consisting of micro-physical and chemical processes (order of 10^{-1} s), temperature variations (order of hours) and seasonal variations (order of months). A family walking in a mall or a park, wherein the parents move at a certain pace while the child moves at a distinctly different pace also constitutes a multiscale system. Multiple timescales can also be induced as a consequence of multirate sampling, that is, different sampling rates for different variables due to sensor limitations and physical constraints on sampling. Note that the phrase *time-scale* is used in a generic sense here. Multiscale nature can be along the spatial dimension or along any other dimension.

Numerical and data-driven analysis of multiscale systems presents serious challenges in every respect, be it the choice of a suitable sampling interval, or the choice of step size in numerical simulation or the design of a controller. The broad appeal and the challenges of these systems have aroused the curiosity of scientists, engineers, mathematicians, physicists, econometricians, and biologists alike. The purpose of this chapter is neither to dwell into the intricacies of multiscale systems nor to present a theoretical analysis of multiscale systems (for recent reviews on these topics, see Braatz et al., 2006; Ricardez-Sandoval, 2011). The objective of this chapter is to present an emerging and an exciting direction in the data-driven analysis of multiscale, time-varying (nonstationary), and nonlinear systems, with focus on empirical modeling (identification) and control. This emerging direction rides on a host of interesting and powerful set of tools arising out of a single transform, namely, the *wavelet transform*. The presentation includes a review of achievements to-date, pointers to gaps in existing works, and suggestions for future work while providing a semi-formal foundation on wavelet theory for beginners.

Applications of wavelet transforms are extremely diverse in functional analysis, analysis of differential equations, signal processing, feature extraction, modeling, monitoring, classification, etc. (see Addison, 2002; Chau et al., 2004; Jaffard et al., 2001). The historical motivation for using wavelet transforms has been to analyze systems (signals) that are *nonstationary*. In a deterministic setting, nonstationary signals are signals with time-varying frequencies, while in a stochastic setting, they are signals whose statistical properties (moments of distribution) change with time. In both cases, however, it is the multiscale behavior of the generating system that is responsible for the nonstationary behavior of the signal. In a broader sense, the term "scale" can be used not only to explain nonstationary characteristics of signals but also to denote the "level" of approximation or resolution in functional analysis, image processing, computer vision, and signal estimation. Several wavelet applications in the literature may not necessarily explicitly stress the multiscale nature of a process as the primary motivation for their use. However, it is implicitly understood that a wavelet-based analysis of a system is warranted only if that system exhibits multiscale (or time-varying) characteristics. This also explains the tone of the introductory paragraphs of this chapter.

Multiscale signals are comprised of components that have different existence times. Certain components have a longer duration, while certain others have a shorter duration. In technical terms, *a multiscale signal comprises components with different time localizations*. On the other hand, multiscale signals also simultaneously possess different *frequency localizations*. An example is that of a musical piece, which consists of different notes (different frequencies) over different time durations. Certain notes persist for a longer period of time, while certain others exist for a short period of time. In engineering applications, measurements are usually made up of contributions from a possibly multiscale process, instrumentation noise, and/or disturbances. Each of these components has a different frequency characteristic and a different settling time. Thus, multiscale analysis of a signal amounts to analyzing its time–frequency-localized characteristics (e.g., amplitude, energy, phase).

Analysis of multiscale signals is also equivalent to constructing multiresolution approximations. For instance, in image processing, each scale corresponds to a *resolution*, a level of fineness or detail. The relation between scale and resolution is vivid in maps of geographical regions where the low scale corresponds to high resolutions (more details) and high scale corresponds to low resolutions (fewer details). Multiresolution approximations are the basis for several image compression and reconstruction algorithms today. An image displayed to the user (e.g., in a browser) is gradually presented at different resolutions starting from the coarsest to the finest possible resolution. These MRAs are facilitated by suitable multiscale tools, wavelets being a popular choice.

In signal processing and control applications, *approximations* of different resolutions result when signals are treated with low-pass filters combined with suitable downsampling operations. Correspondingly, the result of subjecting signals to high-pass filtering operations is the *details*. The ramifications of this correspondence have been tremendous and have led to certain powerful results. *The most remarkable discovery is that of the connections between the multiscale analysis of signals and filtering of signals with a bank of band-pass filters of varying bandwidths.* The gradual discovery of several such connections between time–frequency (T–F) analysis, multiresolution approximations, and multirate filtering brought about a harmonious collaboration of physicists, mathematicians, computer scientists, and engineers, leading to a rapid development of computationally efficient and elegant algorithms for multiscale analysis of signals.

Pedagogically, there exist different starting points for introducing wavelet transforms. In the engineering context, the filtering perspective of wavelets is both a useful and convenient starting point. On the other hand, filters are very well understood and designed in the frequency domain. Therefore, it is natural that multiscale analysis is also connected to a frequency-domain analysis of the system, but at different timescales.

With this motivation, we begin with the T–F approach and gradually expound the filtering connections, briefly passing through the MRA gateway.

Frequency-domain analytic tools, specifically based on the powerful Fourier transform, have been prevalent in every sphere of science and engineering. Spectral analysis, as it is popularly known, reveals valuable process characteristics useful for filter design, signal communication, periodicity detection, controller design, input design (in identification), and a host of other applications. The term spectral analysis is often used to connote Fourier analysis since it essentially involves a frequency-domain breakup of the energy or power (density) of a signal as the case maybe. Interestingly, the seminal work by Fourier, which saw the birth of Fourier series (for periodic signals), was along the *signal decomposition* line of thought in the context of solving differential equations. The work was then extended to accommodate decomposition of finite-energy aperiodic signals. Gradually, by conjoining the Fourier transform with the results by Plancherel and Parseval (see Mallat, 1999), a practically useful interpretation of the transform in the broader framework of *energy/power decomposition* emerged. A key outcome of this synergy is the *periodogram* (Schuster, 1897), a tool that captures the contributions of the individual frequency components of a signal to its overall power. The decomposition of the second-order statistics in the frequency domain was soon found to be a unifying framework for deterministic and stochastic signals through the Wiener–Khintchine theorem (Priestley, 1981), which essentially established a formal connection between the time- and frequency-domain properties. The connection paved way for the spectral representations of stochastic processes, which, in turn, formed the cornerstone for modeling of random processes.

As with every other technique, Fourier transforms and their variants (Proakis and Manolakis, 2005) possess limitations (see Section 3.1 for an illustrated review) in the areas of empirical modeling and analysis. These limitations become grave in the context of multiscale systems. The source of these shortcomings is the lack of any time-domain localization of the Fourier basis functions (sine waves). These basis functions are only suited to capturing the global features of a signal, but not its *local* features. Furthermore, the assumption that a signal is synthesized by amplitude scaled and phase-shifted sine waves is usually more convenient for mathematical purposes than for a physical interpretation. In fact, for all nonstationary signals, there is a complete mismatch between the mathematics of the synthesis and the physics of the process. Thus, Fourier transforms are not ideally suited for multiscale systems, where phenomena are localized in time. In fact, *all single-scale techniques suffer from this limitation, that is, they lack the ability to capture any local behavior of the signal.*

1.2. Historical developments

The problem of extending the frequency-domain analysis to multiscale systems received serious attention from physicists who were interested in developing Fourier-like analysis tools for multiscale systems. The efforts witnessed the birth of *T–F analysis of signals* (Cohen, 1989, 1994).

The two key developments that were contemporaneous and historical to the birth of wavelet transforms were the Short-Time Fourier Transform (STFT) (Gabor, 1946) and Wigner-Ville *distributions* (WVD) (Ville, 1948; Wigner, 1932). Both offered significant improvements over the traditional FT but suffered from shortcomings that severely limited their applicability.

The developments of all T–F analysis tools were based on answers to two critical questions: (i) what choice of basis functions or transforms are ideally suited to the analysis of multiscale systems and (ii) are there fundamental limitations on the ability to localize the energy/power density of a signal in the

T–F plane? An excellent treatment and summary of the historical developments of the subject is given in the books by Cohen (1994) and Mallat (1999). A milestone result is that there exists a fundamental limitation on the ability to localize the energy in the T–F plane given by the well-known *duration–bandwidth principle* (also known under the misnomer *uncertainty principle of signals* citing parallels with Heisenberg's uncertainy principle in quantum physics). The search was then for the "best" transform within the realms of these fundamental limitations. Physicists sought the best T–F atoms, mathematicians searched for the best scale-varying basis functions while the signal processing community hunted for the best bank of multirate band-pass filters.

It was evident that the basis should possess the property of signal under investigation. In the context of multiscale analysis, the requirement was the basis functions should be of windows with finite but different durations. A remarkable contribution was made by Gabor (1946) who brought in a certain degree of time-domain localization to the Fourier transform with the introduction of STFT or Windowed Fourier Transform. The underlying idea was simple—time-localize the signal with a suitable *window function* followed by the usual Fourier transform of the windowed or sliced segment. Gabor's transform could also be thought of analyzing the full-length signal with clipped sine waves. However, the limitations of such an approach were soon realized. The primary issue with this approach is that the frequency span of the clipped basis functions does not adapt to the width of the clip, in accordance to the well-established *duration–bandwidth principle.* Moreover, the choice of window length requires reasonably good *a priori* knowledge of the signal's composition, which calls for trials with different window lengths. Mathematically, the time- and frequency-domain localizations were not elegantly tied to each other. From a signal processing perspective, Gabor's transform was equivalent to subjecting the signal to band-pass filters of fixed bandwidth, not an ideally desirable feature for multiscale analysis.

In the pioneering works by Wigner and Ville, two physicists, a direct decomposition of the energy in the T–F plane was proposed (Ville, 1948; Wigner, 1932). The computation of WVD explicitly avoids the preliminary step of signal transforms, thereby giving certain advantages in terms of the ability to localize the energy in T–F plane. However, a major limitation of the WVD is that the signal is only recoverable up to a phase—a significant limitation in filtering applications.

The historical work of Haar in 1910 (Haar, 1910) presented the first usage of the term *wavelet*, meaning a small (child) wave. Haar, while working in the field of functional analysis, constructed a family of box-like basis functions by scale variation of a single function. The purpose was to achieve multiresolution representations of general functions with multiscale characteristics. The period following Haar's proposition witnessed a spurt of activity on the use of scale-varying basis functions. Paul Levy employed Haar's basis function to investigate Brownian motion where he demonstrated the superiority of Haar wavelet basis to Fourier basis in studying short-lived complicated details (Meyer, 1992).

Three decades later, Weiss and Coifman (1977) studied basis functions, termed as *atoms* for T–F analysis of signals. Nearly two decades later, the combined work of Grossmann and Morlet (1984) formalized the theory of wavelets and wavelet transforms. Morlet's findings (Morlet et al., 1982) stemmed from his efforts to analyze seismic signals of different durations and frequencies as an engineer, while Grossman's results originated from his efforts to find suitable T–F atoms in the context of quantum physics. The original wavelet transform is a *redundant* or a dense transform, meaning that it required more bases than necessary to decompose a signal in the T–F plane. Meyer's works (Meyer, 1985, 1986) opened gateways into *orthogonal* wavelet transforms, which have attractive properties, mainly that of a minimal representation of a signal with good T–F localization. Shortly thereafter, the discovery of the remarkable connections between orthogonal wavelet bases and quadrature mirror filters in signal processing (Mallat, 1989b) provided a big impetus to the world of wavelets, in the same way as Cooley–Tukey's fast Fourier transform (FFT) algorithm (Proakis and Manolakis, 2005). Mallat (1989b) showed that the decomposition of signal onto orthogonal wavelet bases at different scales can be efficiently implemented by a multistage pyramidal algorithm consisting of a cascaded low-pass, high-pass filtering operations combined with downsampling operations at every stage.

The connections between multiresolution approximations and orthonormal wavelet bases (Mallat, 1989a), signal processing and wavelet bases (Mallat, 1989b) essentially established that the MRA can be achieved by the design of special filter banks known as *conjugate mirror filters* (Vaidyanathan, 1987; Vetterli, 1986). Conditions on bases could be translated to appropriate constraints on filters. In T–F analysis, wavelets were shown to offer an adaptive trade-off between the time and frequency localizations of the wavelet atoms. The adaptivity is not with respect to the signal *per se*, but with respect to the frequency band under scrutiny. Low-frequency components are analyzed using wide windows, while high-frequency components are analyzed using narrow windows (good time localization) in abeyance with the duration–bandwidth principle. Wavelet filters thus provide a constant relative bandwidth unlike STFT which offers bank of filters with constant bandwidth (Cohen, 1994). Another aspect in which wavelet transform outscores STFT and traditional filtering methods is that the entire family of band-pass filters can be merely condensed to the design of a single filter. In addition, they are excellent at providing sparse representations of a wide class of signals. Equipped with several attractive properties, wavelets soon found an indispensable place in a diverse set of applications such as signal compression, estimation (denoising), T–F analysis, feature extraction, multiscale modeling, and monitoring of process systems.

The literature of wavelet transforms today is inundated with numerous variants of wavelet transforms and their implementations, each tailored for a specific end-use. All such variants of transforms are based on a single wavelet transform, which is the continuous wavelet transform (CWT). Innumerable tutorial/research articles in several dedicated journals, excellent textbooks on foundations and application aspects of wavelet transforms and open source web-based course material bear testimony to the enormous utility of wavelet transforms (Addison, 2002; Chau et al., 2004; Chui, 1992; Gao and Yan, 2010; Jaffard et al., 2001; Mallat, 1999; Motard and Joseph, 1994; Percival and Walden, 2000). A list of free and commercial wavelet software packages is found in Lio (2003). The list is left incomplete without the mention of the T–F toolbox (Auger et al., 1997) and the WTC toolbox (Grinsted et al., 2002). Wavelet transforms have inspired several researchers to propose new transforms or innovate existing transforms. Examples of such developments are ridgelet, curvelet, and contourlet transforms (AlZubi et al., 2011; Candes and Donoho, 1999, 2000; Do and Vetterli, 2005; Ma and Plonka, 2010).

Wavelets have been used in different forms in modeling, control, and monitoring of processes depending on the requirement. They have offered immense benefits in a multitude of process systems applications—signal/data compression, data preprocessing and data reconciliation, signal estimation/filtering, a basis for signal representation for multiscale systems, process monitoring, and feature extraction. They have also been used in deriving solutions to partial differential equations with limited applicability. In engineering applications, wavelets have been used in two different ways: (i) as preprocessing tools and (ii) in integration with other single-scale univariate/multivarite method.

The prime objectives of this chapter are (i) to provide a tutorial introduction to wavelet transforms that facilitates easy understanding of the subject,

(ii) to present an overview of applications and the relevant concepts of wavelet transforms in analysis of multiscale systems, and (iii) to present new ideas for identification of multiscale systems using spline biorthogonal wavelets as basis.

1.3. Outline

The organization of this chapter is as follows. Section 2 presents the connections between the world of transforms, approximations, and filtering with the intention of enabling the reader to smoothly connect the different birth points of wavelets. Practically the subject of Fourier transforms is considered as a good starting point in understanding wavelet theory. Justifiably Section 3.1 reviews Fourier transforms and their properties. This is followed by Section 3.3, which presents a brief review of the STFT and WVD, the two major developments en route to the emergence of wavelet transforms.

Section 4 introduces wavelet transforms to the reader with focus on continuous- and discrete wavelet transform (CWT and DWT), the two most widely used forms of wavelet transforms. The connections between multiresolution approximations, T–F analysis, and filtering are demonstrated. A brief discussion on variants of these transforms is included.

In Section 6, we present an in-depth review of applications to modeling (identification) and control (design and performance assessment). Signal estimation and achieving sparse representations are key steps in modeling. Therefore, applications to signal estimation are reviewed in Section 5 as a precursor. Particular attention is drawn to the less-known, but very effective, concept of consistent estimation with wavelets.

In Section 7, an alternative identification methodology using wavelets is put forth. The key idea is to develop models in the coefficient domain using the idea of consistent prediction (stemming from consistent estimation concepts). Applications to simulation case studies and an industrial application are presented.

The chapter concludes in Section 8 offering closing remarks and ideas that merit exploration.

2. TRANSFORMS, APPROXIMATIONS, AND FILTERING

In the discussions to follow, the signal is treated as a function denoted by $x(t)$ or $f(t)$ (continuous-time) or as a vector of samples x (discrete-time case) depending on the context.

2.1. Transforms

Transforms are frequently used in mathematical analysis of signals to study and unravel characteristics that are otherwise difficult to discover in the raw domain. Any signal transformation is essentially a change of representation of the signal. A sequence of numbers in the original domain is represented in another domain by choosing a different basis of representation (much alike in choosing different units for representing weight, volume, pressure, etc.). The expectation is, that in the new basis, certain features (of the signal) of interest are significantly highlighted in comparison to the original domain where they remain obscure or hidden due to either the choice of original basis or the presence of measurement noise. It is to be remembered that a *change of basis can never produce new information, but only the way in which information is represented or captured.*

The choice of basis clearly depends on the features or characteristics we wish to study, which is in turn driven by the application. On the other hand, the new basis should satisfy an important requirement *of stability*, that is, the new "numbers" do not become unbounded or divergent. Moreover, in several applications, it may be additionally required to uniquely recover the original signal from its transform, that is, the transform should not result in loss of information and should be without ambiguity.

Interesting perspectives of transforms emerge when one views a transform as projections onto basis functions and/or a filtering operation. The choice/design/implementation of a transform then amounts to choosing/designing a particular set of basis functions followed by projections or from a signal processing perspective, the choice/design/implementation of a filter.

In data analysis, Fourier transform is used whenever it is desired to investigate the presence of oscillatory components. It involves projection/correlation of the signal with sinusoidal basis and is stable only under certain conditions, while guaranteeing perfect recovery of the signal whenever the transform exists.

From the foregoing discussion, it is clear that transformation of a signal is equivalent to representing the signal in a new basis space. The transform itself is contained in the projection or the shadow of the given signal onto the new basis functions.

2.2. Projections and projection coefficients

Working in a transform domain amounts to analyzing a measurement using its projection coefficients $\{c_i\}$ *rather than the measurement or its projections* because the coefficients usually enjoy certain desirable features and statistical properties that are not possessed by either the measurement or its projections.

A classic example is the case of a sine wave embedded in noise. A sine wave embedded in noise is difficult to detect by a mere visual inspection of the measurement in time-domain. However, a Fourier transform (projection) of the signal produces coefficients that facilitate excellent separation between the signal and noise. A pure sine wave produces very few nonzero high-amplitude coefficients in the Fourier basis space, while the projections of noise yield several low to very low amplitude coefficients. Thus, the separation of sine wave is greatly enhanced in the transform space.

Another example is that of the DWT of a signal that exhibits significant intrasample correlation. The autocorrelation is broken up by the DWT to produce highly decorrelated coefficients. This is a useful property explored in several applications.

In addition to separability and decorrelating ability, sparsity is a highly desirable property of a transform (e.g., in signal compression, modeling). In the sine wave example, the signal has a *sparse* representation in the Fourier domain. Wavelet transforms are known to produce sparse representations of a wide class of signals.

The three preceding properties of a transform (projection) render transform techniques indispensable to estimation. Returning to the sine wave example, when the objective is to recover (estimate) the signal, one can reconstruct the signal from its projections onto the select basis (highlighted by peaks in the coefficient amplitudes) alone, that is, the projections onto other basis functions are set to zero. This is the principle underlying the popular Wiener filter (Orfanidis, 2007) for signal estimation and all thresholding algorithms in the estimation of signals using DWT.

Separation of a signal into its approximation and detail constituents is the central concept in all filtering and estimation methods. In signal estimation, approximations of measurements are constructed to extract the underlying signal. The associated residuals carry the left out details, ideally containing undersirable portions, that is, noise.

2.3. Filtering

The foregoing observations bring out a synergistic connection between the operations of filtering, projections, and transforms. Qualitatively speaking, approximations are *smoothed* versions of $x(t)$. The details should then naturally contain the *fluctuating* portions of $x(t)$. In filtering terminology, approximations and details are the outputs of the low-pass and high-pass filters acting on $x(t)$.

Filtering applications of transforms are best understood and implemented when the transform basis set is a family of orthogonal vectors. With an orthogonal basis set, details are termed as *orthogonal complements* of the approximations. Mathematically, the space spanned by the details is orthogonal to the space spanned by the approximations. This is the case with both Fourier Transforms and Discrete Wavelet Transforms.

Transform of a signal can also be written as its convolution with the basis function of the transform domain. From systems theory, convolution operations are essentially filtering operations and are characterized by the impulse response (IR) functions of the associated filters. For example, the STFT and the Wavelet Transform can be written as convolutions that bring out their filtering nature.

2.4. Correlation: Unified perspective

Appendix A shows that transforms or projections essentially involve inner products of the signal with the transform basis. Inner products are measures of similarity. The correlation[1] (in a signal processing sense) between two signals (functions) $f(t)$ and $g(t)$ in an inner product space is defined as

$$\mathrm{corr}(f(t), g(t)) = \langle f(t), g(t) \rangle = \int_{-\infty}^{\infty} f(t) g^*(t) \mathrm{d}t$$

Transforms therefore work with correlations; similarly, projection coefficients are correlations. It follows that filtering is also a correlation operation. All of them measure similarity between the signal and the basis function. The point that calls for a reiteration is that the choice of basis function is dependent on what we wish to detect in or extract from the signal.

3. FOUNDATIONS

3.1. Fourier basis and transforms

Fourier Transform is perhaps one of the most widely used ubiquitous transform in signal processing and data analysis. It also occupies a prominent place in all spheres of engineering, mathematics, and sciences. This transform mobilizes sines and cosines as its basic vehicles.

[1] Correlation in statistics is defined differently—it is the normalized covariance.

The origins of Fourier transform trace back to Fourier's proposition of solving heat wave equations using a series expansion of the solution in sines and cosines (Fourier, 1822, also see Bracewell, 1999). In due course of its adaptation, the transform acquired different names depending on the nature of the signal, that is, whether it is periodic/aperiodic or continuous/discrete in the original domain (usually time domain) (see Proakis and Manolakis, 2005 for an excellent exposition).

Among the variants, the discrete-time Fourier transform and its finite-sample version, discrete FT are most relevant:

$$X(f) \triangleq \sum_{k=-\infty}^{\infty} x[k] e^{-j2\pi fk} (\text{analysis}) \quad x[k] = \int_{-1/2}^{1/2} X(f) e^{jfk} \mathrm{d}f \, (\text{synthesis}) \quad [3.1]$$

$$X[f_n] = X[n] \triangleq \sum_{k=0}^{N-1} x[k] e^{-j2\pi kn/N} (\text{analysis}) \quad x[k] = \frac{1}{N} \sum_{n=0}^{N-1} X[n] e^{j2\pi kn/N} (\text{synthesis})$$
$$f_n = \frac{n}{N}, n = 0,1,\ldots,N-1 \qquad k = 0,1,\ldots,N-1 \quad [3.2]$$

The forward transform is also known as the *analysis* or *decomposition* expression, while the inverse transform is known as the *synthesis* or *reconstruction* expression. Interestingly, the inverse transform is usually the starting point of a pedagogical presentation. The analysis equation provides the projection coefficients of the corresponding projection. These coefficients are *complex valued* in general.

For computational purposes, an efficient algorithm known as the FFT algorithm is used. The interested reader is referred to Proakis and Manolakis (2005) and Smith (1997) for implementation details and Cooley et al. (1967) for a good historical perspective.

3.1.1 Fourier coefficients

The Fourier projection coefficients possess certain extremely useful properties and interpretations. Operations in time-domain transform to operations on the coefficients in frequency domain. Several standard texts on signal processing discuss these properties in detail (see Oppenheim and Schafer, 1987; Proakis and Manolakis, 2005). Three properties relevant to the understanding of wavelet transforms are highlighted below.

i. *Convolution* operation in the signal space transforms to a *product* in the coefficient space

$$x_3(t) = x_1 * x_2(t) \triangleq \int_{-\infty}^{\infty} x_1(\tau)x_2(t-\tau)\mathrm{d}\tau \xrightarrow{\mathcal{F}} X_3(\omega) = X_1(\omega)X_2(\omega) \tag{3.3}$$

This is a remarkably useful result in theoretical analysis of signals and systems

ii. Parseval's result (energy preservation)

$$E_{xx} = \sum_{k=-\infty}^{\infty} |x[k]|^2 = \int_{-1/2}^{1/2} |X(f)|^2 \mathrm{d}f \tag{3.4}$$

The squared amplitude of the coefficients, $|X(f)|^2$ or $|X(f_n)|^2$ as the case may be, thus qualify to be the energy density or power distribution of the signal in frequency domain. Thus, *a signal decomposition is actually a spectral decomposition of the power/energy.*

iii. Time-scaling property:

$$\text{If } x_1(t) \xrightarrow{\mathcal{F}} X_1(\omega) \text{ then } \frac{1}{\sqrt{s}} x_1\left(\frac{t}{s}\right) \xrightarrow{\mathcal{F}} X_1(s\omega) \tag{3.5}$$

If $x_1(t)$ is such that $X_1(\omega)$ is centered around ω_0, then *time*-scaling the signal by s shifts the center of $X_1(\omega)$ to ω_0/s. A very useful property in understanding the equivalence between scaling in wavelet transforms and their filtering abilities.

3.1.2 Limitations of Fourier analysis

The reign of Fourier transforms is supreme in the world of signals that are stationary, that is, signals consisting of same frequencies at all times. However, its application to signals which are made up of different frequency components over different time intervals is very limited. This should not be construed as a mathematical limitation of Fourier transforms, but rather as its *unsuitability* for such signals.

The prime reason is the infinite time-spread (zero time localization) of the FT basis functions limiting their ability to extract only the *global* and not the *local* (temporal) oscillatory features of a signal. Furthermore, these basis functions force the transform to represent zero-activity time-regions of a signal as additions and cancelations of sine waves, which is mathematically perfect, but a far cry from the physics of the signal-generating process.

Changes in time-domain features are indeed contained in the phase $\angle X(f)$, which is why perfect reconstruction of the signal is possible. However, extracting the time-varying frequency content from phase is a highly intractable task complicated by certain limitations. Moreover, estimation of phase is very sensitive to the presence of noise.

The shortcomings are illustrated by means of two examples.

Example 1: The first example is that of two measurements containing identical frequencies $f_1 = 0.06$ and $f_2 = 0.16$ cycles/sample, but over different time intervals. Figure 3.1A shows the spectral densities for these two different signals. Clearly spectral density is invariant to the time localization of the frequency components.

Example 2: The second example is concerned with two signals consisting of a sine wave of frequency $f = 0.05$ cycles/sample corrupted with an impulse at two different instants $k = 105$ and 145 instants. From the spectral density shown in Fig. 3.1B, there is no means for determining the location of the impulse.

In both examples, one could ideally use the phase for determining the time stamps of frequencies, but its applicability is very limited due to the complicated behavior of phase in presence of noise and when the same band of frequencies are present over different intervals.

Turning to methods that present energy/power spectral densities in the joint T–F plane, a question that naturally emerges is whether one can capture the localization of energy density in time and frequency domains simultaneously with arbitrary accuracy. Unfortunately, the answer is a no. This is due to a standard result in signal processing, known as the *duration–bandwidth* principle, which is reviewed below. An excellent treatment of this result with proper interpretations is given in the text by Cohen (1994), which also inspires the presentation of the ensuing section.

3.2. Duration–bandwidth result

The main result is stated below. The proof is found in many standard texts (see Cohen, 1994; Oppenheim and Schafer, 1987).

The energy spread of a signal $x(t)$ in time measured by the duration σ_t^2 and energy spread in frequency of its Fourier transform $X(\omega)$ measured by the bandwidth σ_ω^2 necessarily satisfy[2]

$$\sigma_t^2 \sigma_\omega^2 \geq \frac{1}{4} \qquad [3.6]$$

[2] A rigorous lower bound is derived in Cohen (1994).

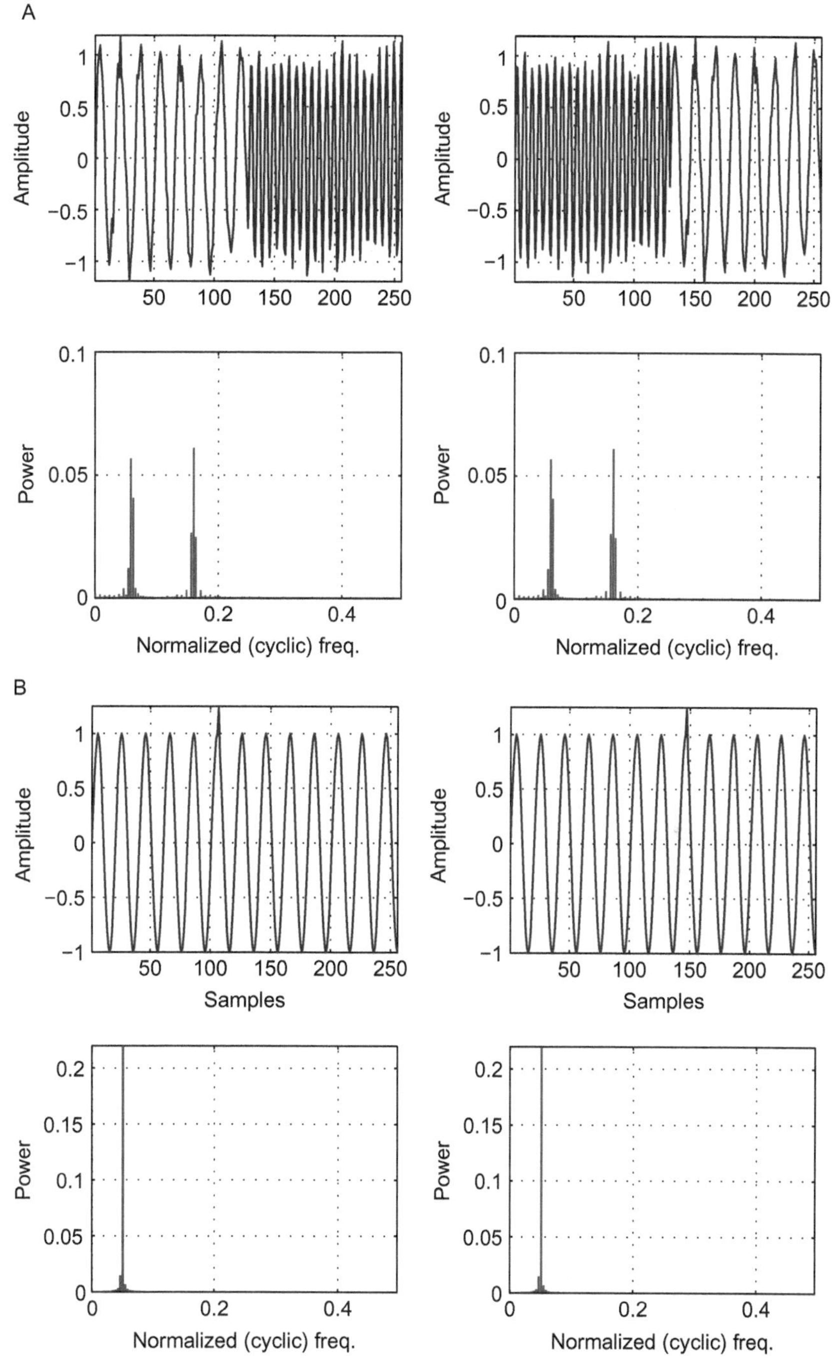

Figure 3.1 FT is insensitive to time-shifts of frequencies or components in a signal. (A) Frequencies are reversed in time and (B) impulses at different times. (For color version of this figure, the reader is referred to the online version of this chapter.)

Remarks

1. The quantities σ_t^2 and σ_ω^2 are defined as

$$\sigma_t^2 = \int_{-\infty}^{\infty} (t - \langle t \rangle)^2 |x(t)|^2 \, dt = \langle t^2 \rangle - \langle t \rangle^2 \quad [3.7]$$

$$\sigma_\omega^2 = \int_{-\infty}^{\infty} (\omega - \langle \omega \rangle)^2 |X(\omega)|^2 d\omega = \langle \omega^2 \rangle - \langle \omega \rangle^2 \quad [3.8]$$

   where $\langle t \rangle$ and $\langle \omega \rangle$ are the averages time and frequency, respectively, as measured by the energy densities $|x(t)|^2$ and $|X(\omega)|^2$, respectively.
2. The duration and bandwidth are second-order central moments of the energy densities in time and frequency, respectively (analogous to the statistical definition of variance).
3. The result is only valid when the density functions are a Fourier transform pair.

Equation (3.6) is reminiscent of the uncertainty principle due to Heisenberg in quantum mechanics, which is set in a probabilistic framework and dictates that the position and momentum of a particle cannot be known simultaneously with arbitrary accuracy. Owing to this resemblance, Eq. (3.6) is popularly known as the *uncertainty principle for signals*. However, the reader is cautioned against several prevailing misinterpretations. Common among them are that time and frequency cannot be made arbitrarily narrow, time and frequency resolutions are tied together and so on.

The consequence of the duration–bandwidth principle is that, using Fourier transform-based methods, *it is not possible to localize the energy densities in time and frequency to a point in the T–F plane*. In passing, it should be noted that when working with the joint energy density in the T–F plane, two duration–bandwidth principles apply. The first one involves the local quantities (duration of a given frequency ω and bandwidth at a given time t), while the other is based on the global quantities. The limits on both these products have to be rederived for every method that constructs the joint energy density.

3.3. Short-time transitions

Within the boundaries imposed by the duration–bandwidth principle, one can still significantly segregate the multiple time-scale components of a signal and localize the energy densities within a T–F cell (tile). The difference in various T–F tools is essentially in nature of the tiling of the energy densities in the T–F plane.

The Windowed Fourier Transform, also known as the STFT proposed by Gabor (1946), was among the first ones to appear on the arena. The idea is intuitive and simple. Slice the signal into different segments (with possible overlaps) and subject each slice to a Fourier transform. The slicing operation is equivalent to windowing the signal with a *window function* $w(t)$.

$$x(t_c, t) = x(t)w(t - t_c) \qquad [3.9]$$

where t_c denotes the *center* of the window function. The window function is naturally required to satisfy an important requirement, that of the *compact support*.

Compact support: The window $w(t)$ (with $W(\omega)$ as its FT) should decay in such a way that

$$x(t_c, t) = \begin{cases} x(t)w(t - t_c) & \text{for } t \text{ near } t_c \\ 0 & \text{for } t \text{ far away from } t_c \end{cases}$$

and have a length shorter than the signal length for the STFT to be useful.

In addition, a unit energy constraint $\| w \|_2^2 = 1$ is imposed to preserve the energy of the sliced signal.

The STFT is the Fourier transform of the windowed signal,

$$X(t_c, f) = \int_{-\infty}^{\infty} x(t_c, t)e^{-j\omega t}\,dt = \int_{-\infty}^{\infty} x(t)w(t - t_c)e^{-j\omega t}\,dt \qquad [3.10]$$

An alternative viewpoint is that STFT is the transform of the signal $x(t)$ with clipped sinusoids $w(t - t_c)e^{-j\omega t}$ as basis functions. This viewpoint explains the improvement brought about by STFT over the FT by highlighting that it uses basis functions that have compact support.

The energy decomposition of the signal achieved by STFT in the T–F plane is given by

$$P(t_c, \omega) = |X(t_c, \omega)|^2 = \left| \int_{-\infty}^{\infty} x(t)w(t - t_c)e^{-j\omega t}\,dt \right|^2 \qquad [3.11]$$

The *spectrogram* $P(t_c,\omega)$ is the energy density in the T–F plane due to the fact that

$$\int_{-\infty}^{\infty} |x(t)|^2 dt = \frac{1}{2\pi}\int_{-\infty}^{\infty} |X(\omega)|^2 d\omega = \frac{1}{2\pi}\int_{-\infty}^{\infty}\int_{-\infty}^{\infty} P(t_c, \omega)\,d\omega\,dt_c \qquad [3.12]$$

The discrete STFT (also known as the Gabor transform) is given by

$$X[m, l] = \langle x[k], \gamma[m, l; k] \rangle = \sum_{k=0}^{N-1} x[k]h[k - m]e^{(-j2\pi lk/m)} \qquad [3.13]$$

where $\gamma[m, l; k] = h[k-m]e^{j(2\pi lk/m)}$ is the discrete windowed Fourier basis or atom. Increments in the center of window denoted by m during the analysis determine whether one achieves a redundant (overlapping windows) or an orthogonal representation.

The STFT, like its predecessor, enjoys certain useful properties while at the same time suffering from limitations. To keep the discussion focussed, we review only the important ones.

i. *Filtering perspective:*

$$\begin{aligned} X(t_c, \omega_0) &= \int_{-\infty}^{\infty} x(t)w(t-t_c)e^{-j\omega_0 t}dt \\ &= e^{-j\omega_0 t_c}\int_{-\infty}^{\infty} x(t)w(t_c-t)e^{j\omega_0(t_c-t)}dt \end{aligned} \quad [3.14]$$

where we have used the symmetry property $w(t) = w(-t)$. The integral in Eq. (3.14) is a convolution, meaning the STFT at (t_c, ω_0) is $x(t)$ filtered by $W(\omega-\omega_0)$, which is a band-pass filter whose bandwidth is governed by the time-spread of $w(t)$. The quantity $e^{-j\omega_0 t_c}$ is simply a modulating factor and results only in a frequency shift. Thus, STFT is equal to the result of passing the signal through a band-pass filter of constant bandwidth.

ii. *T–F localization:* Two test signals are used to evaluate the localization properties

$$x(t) = \delta(t-t_0) : X(t_c, \omega) = w(t_0 - t_c)e^{-j2\pi\omega t_0} \Rightarrow P(t_c, \omega) = |w(t_0 - t_c)|^2 \quad [3.15]$$

$$x(t) = e^{j2\pi\omega_0 t} : X(t_c, \omega) = W(\omega-\omega_0)e^{-j2\pi\omega_0 t_c} \Rightarrow P(t_c, \omega) = |W(\omega-\omega_0)|^2 \quad [3.16]$$

Thus, the time and frequency localizations of the energy/power density are completely determined by the energy spreads of the window function in the respective domains.

A narrow window in time produces very good energy localization in time, but by virtue of the limitation in Eq. (3.6) produces a large smearing of energy in frequency domain. The same argument applies to a narrow window in frequency domain. It produces large smearing of energy in time domain. It is instructive to verify that when $w(t) = 1$, $-\infty < t < \infty$, STFT reduces to FT, completely losing its ability to localize the energy in time.

iii. *Window type and length:* Eqs. (3.15) and (3.16) indicate that both the window type and length characterize the behavior of STFT. Several choices of window functions exist (Proakis and Manolakis, 2005). A suitable one is that offers a good trade-off between edge effects (due to finite length) and resolution. Popular choices are Hamming, Hanning, and Kaiser windows (Proakis and Manolakis, 2005).

The window length plays a crucial role in localization. Figure 3.2 illustrates the impact of window lengths on the spectrogram for a signal $x[k] = \sin(2\pi 0.15k) + \delta[k-100]$, where $\delta[.]$ is the Kronecker delta function. The narrower window is able to detect the presence of the small disturbance in the signal but loses out on the frequency localization of the sine component. Observe that the Fourier spectrum is excellent at detecting the sine wave, while it is extremely poor at detecting the presence of the impulse.

The preceding example is representative of the practical limitations of STFT in analyzing real-life signals. The decision on the "optimal" window length for a given situation rests on an iterative approach to be adopted by the user.

The STFT is accompanied by two major shortcomings:

- The user has to select an appropriate window length (that detects both time- and frequency-localized events) by trial and error. This involves a fair amount of book keeping and a compromise (of localizations in the T–F plane) that is not systematically achieved.
- A wide window is suitable for detecting long-lived, low-frequency components, while a narrow window is suitable for detecting short-lived, high-frequency components. The STFT does not tie these facts together and performs a Fourier transform over the entire frequency range of the segmented portion.

Figure 3.3 illustrates the benefits and shortcomings of the STFT in relation to the FT.

A transform that ties the tiling of the T–F axis in accordance with the duration–bandwidth principle is desirable. From a filtering viewpoint, choosing a wide window should be tied to low-pass filtering while a narrow window should be accompanied by high-pass filtering. Thus, *the key is to couple the filtering nature of a transform with the window length.* Wavelet transforms were essentially built on this idea using the scaling parameter as a coupling factor.

3.4. Wigner–Ville distributions

Prior to the emergence of transforms that could facilitate T–F localization, Wigner (1932) and Ville (1948) independently laid down ideas for methods

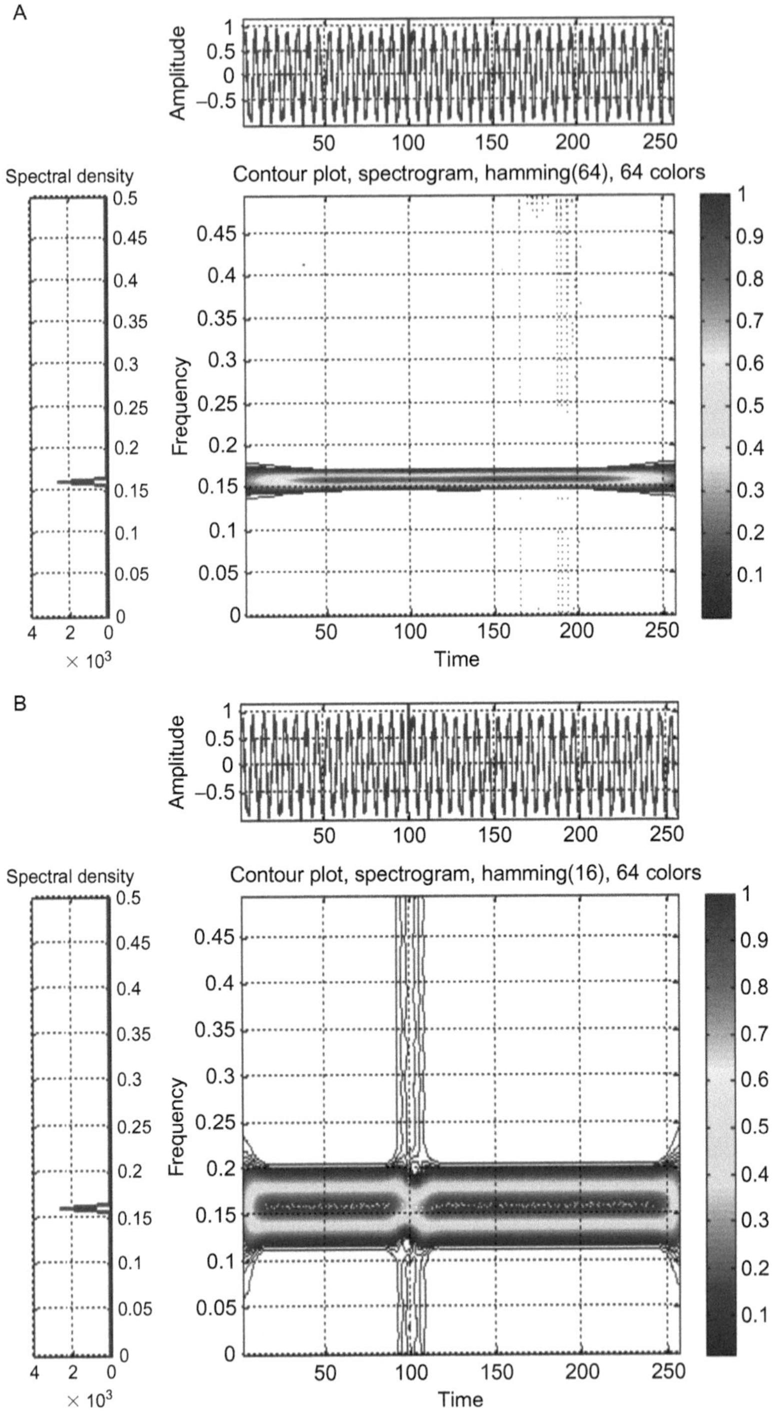

Figure 3.2 Spectrogram of a test signal (sine wave corrupted by an impulse) with two different window lengths, $L_1 = 64$ and $L_2 = 16$ samples. (A) Hamming window of length 64 samples and (B) Hamming window of length 16 samples. (For color version of this figure, the reader is referred to the online version of this chapter.)

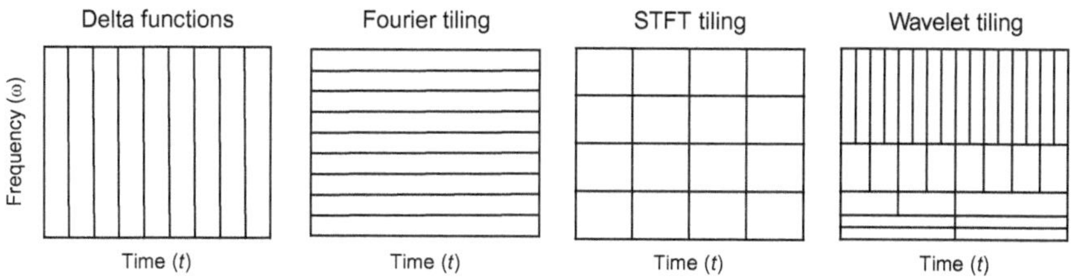

Figure 3.3 Tiling of the T–F plane by the time-domain sampling, FT, STFT, and DWT basis.

that avoided the transform route by directly computing the joint energy density function from the signal. The result was the WVD (Cohen, 1994; Mallat, 1999), which provided excellent T–F localization of energy. Mathematically, the distribution is computed as

$$\begin{aligned} WV(t,\omega) &= \frac{1}{2\pi}\int x^*\left(t-\frac{\tau}{2}\right)x\left(t+\frac{\tau}{2}\right)\mathrm{e}^{-j\tau\omega}\mathrm{d}\tau \\ &= \frac{1}{2\pi}\int X^*\left(\omega-\frac{\theta}{2}\right)X\left(\omega+\frac{\theta}{2}\right)\mathrm{e}^{-j\theta\omega}\mathrm{d}\theta \end{aligned} \quad [3.17]$$

The WVD satisfies several desirable properties of a joint energy distribution function such as shift invariance, marginality conditions (unlike the STFT), finite support, etc., but suffers from a few critical shortcomings (see Cohen, 1994).

i. *$WV(t, \omega)$ is not guaranteed to be positive valued.* This is a crucial drawback.

ii. WVD expresses the energy of a signal as a sum of the energies of individual components plus interference terms, which are spurious artifacts (Mark, 1970).

Subsequent efforts to produce a positive-valued distribution function and to alleviate the interference artifacts resulted in convolutions of the WVD in Eq. (3.17) with a smoothing *kernel* (Claasen and Mecklenbrauker, 1980). These are known as the pseudo- and smoothed-WVDs. The Cohen's class of functions (Cohen, 1966) offers a unified framework for all such smoothed WVD methods. Figure 3.4 illustrates the interference terms introduced by WVD for a composite signal and the subsequent removal of the same by a pseudosmoothed WVD, however at the expense of loosing the fine localization achieved by WVD.

What also followed subsequently was a fascinating equivalence result—the *spectrogram and scalogram (wavelet-based) are essentially smoothed*

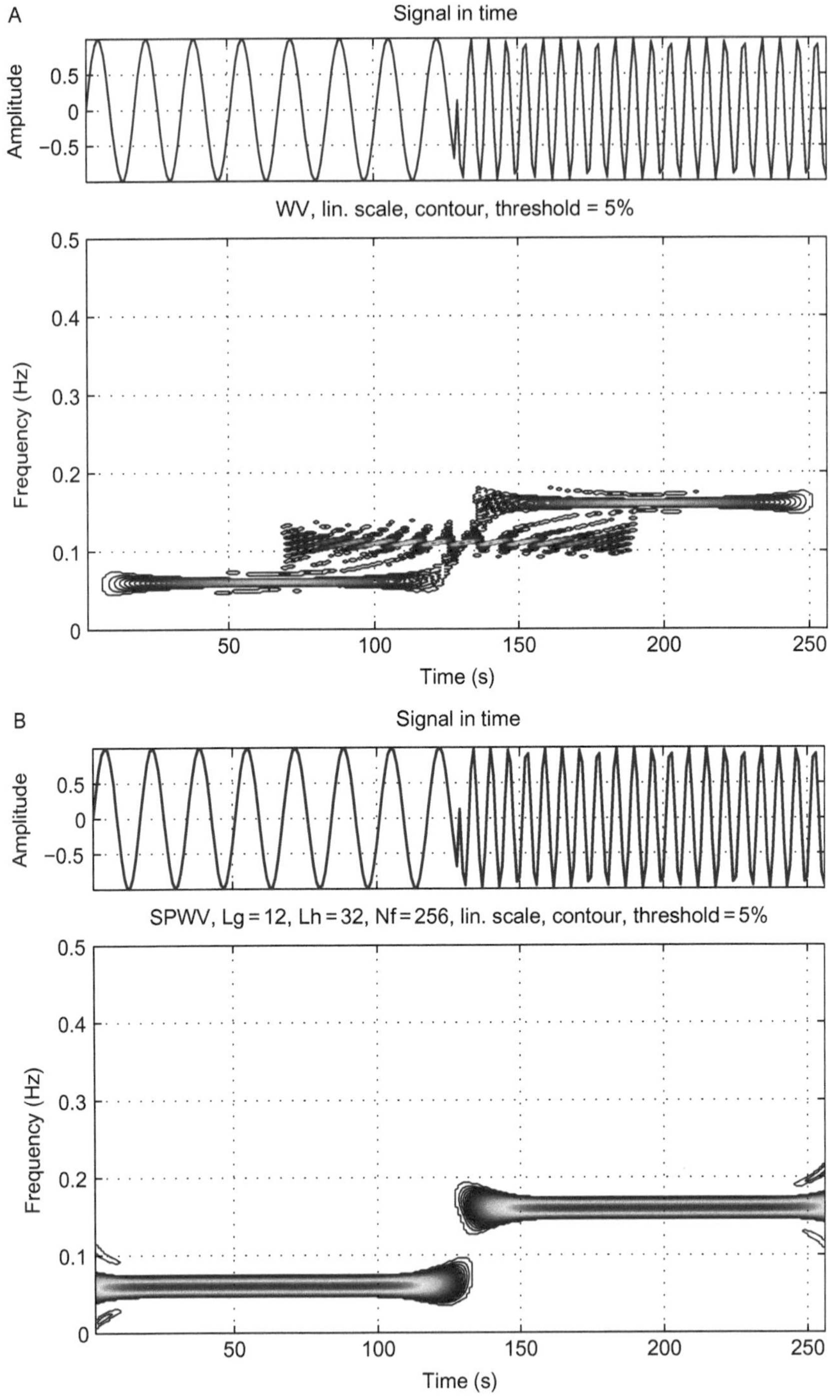

Figure 3.4 Artifacts introduced by WVD are eliminated by a suitable smoothing—at the expense of localization. (A) Wigner-Ville distribution and (B) pseudosmoothed WVD. (For color version of this figure, the reader is referred to the online version of this chapter.)

WVDs with different kernels (Cohen, 1994; Mallat, 1999; Mark, 1970). It is also possible to start from the spectrogram or scalogram and arrive at WVD by an appropriate smoothing.

An interesting consequence of smoothing the WVD is that while it guaranteed positive-valued functions and eliminated interferences, the marginality condition was lost. This was not surprising though due to Wigner's own result which stated that *there is no positive quadratic energy distribution that satisfies the time and frequency marginals* (see Wigner, 1971).

iii. Signal cannot be recovered unambiguously from its WVD since the phase information required for perfect reconstruction is lost. This is akin to the fact that it is not possible to recover a signal from its spectrum alone. Thus, WVD and its variants are not the ideal tools for filtering applications.

Notwithstanding the limitations, pseudo- and smoothed-WVDs offer tremendous scope for applications primarily due to their good energy density localization (e.g., see Boashash, 1992). With this historical perspective, it is hoped that the reader will develop an appreciation of the wavelet transforms and place it in proper perspective.

4. WAVELET BASIS, TRANSFORMS, AND FILTERS

The idea behind wavelet transforms is similar to that of performing several STFT analyses with different window sizes, but in an intelligent manner. For a given window length, the bandwidth of the STFT filter is constant for the entire ω-axis. On the other hand, wavelet windows of different lengths are coupled with their filtering abilities in accordance with the duration–bandwidth principle—wide windows search for long-lived low frequencies, while narrow windows search for short-lived low frequencies. Thus, *wide wavelets have a narrowband frequency response with center frequency in the low-frequency regime and narrow wavelets have a broadband frequency response, but centered in the high-frequency zone.*

These ideas are realized by *scaling* and *translating* only one basis function known as the mother wave. This is the first basic difference between a wavelet transform and STFT. The *finite energy* "mother" wavelet function $\psi(t)$ should possess a zero mean,

$$\int_{-\infty}^{\infty} \psi(t)\mathrm{d}t = 0 = \hat{\psi}(\omega)|_{\omega=0} \qquad [3.18]$$

which can also be conceived as a requirement for the wavelet to act as a band-pass filter.[3] In (3.18), the hat on ψ indicates that it is a Fourier transformed quantity.

The family of wavelets is generated by scaling and translating the mother wave,

$$\psi_{\tau,s} = \frac{1}{\sqrt{|s|}}\psi\left(\frac{t-\tau}{s}\right), \quad \tau, s \in R \; s \neq 0 \qquad [3.19]$$

where $\tau \equiv t_c$ (in STFT) is the *translation* parameter used to traverse along the length of the signal. The factor $1/\sqrt{|s|}$ is a normalization factor to ensure $\|\psi_{t,s}\|_2 = \|\psi\|_2$.

The *scaling* parameter s determines the compression or dilation of the mother wave. If $s > 1$, $\psi_{\tau,s}(t)$ is in a dilated state, resulting in a wide window or equivalently a low-pass filter. On the other hand, if $0 < s < 1$, $\psi_{\tau,s}(t)$ is in a compressed state, producing narrow windows that are suitable for analyzing the high-frequency components of the signal.

4.1. Continuous wavelet transform

The CWT of a function $x(t)$ is its coefficient of projection onto the wavelet basis (Grossmann and Morlet, 1984; Jaffard et al., 2001; Mallat, 1999),

$$Wx(\tau,s) = \frac{\left\langle x, \psi_{\tau,s}\right\rangle}{\|\psi_{\tau,s}\|_2^2} = \int_{-\infty}^{+\infty} x(t)\psi^*_{\tau,s}(t)\,\mathrm{d}t \qquad [3.20]$$

Thus, CWT is the correlation between $x(t)$ and the wavelet dilated to a scale factor s but centered at τ.

As in FT, the original signal $x(t)$ can be restored perfectly using

$$f(t) = \frac{1}{C_\psi}\int_0^{\infty}\int_{-\infty}^{+\infty} Wx(\tau,s)\psi_{\tau,s}\frac{1}{s^2}\mathrm{d}s\,\mathrm{d}\tau = \frac{1}{C_\psi}\int_0^{\infty} Wx(.,s) * \psi_s(t)\frac{\mathrm{d}s}{s^2}, \qquad [3.21]$$

provided the condition on *admissibility constant*

$$C_\psi = \int_0^{\infty}\frac{\hat{\psi}^*(\omega)\hat{\psi}(\omega)}{\omega}\mathrm{d}\omega < \infty \qquad [3.22]$$

is satisfied. This is guaranteed as long as the zero-average condition (3.18) is satisfied.

[3] Note that $\psi(t)$ is not necessarily symmetric unlike in STFT.

Energy preservation: Energy is conserved according to

$$\int_{-\infty}^{\infty} |x(t)|^2 \mathrm{d}t = \frac{1}{C_\psi} \int_0^{\infty} \int_{-\infty}^{\infty} |Wx(\tau,s)|^2 \mathrm{d}\tau \frac{\mathrm{d}s}{s^2} \quad [3.23]$$

4.1.1 Understanding the scale parameter

Given that wavelets emerged largely from the fields of mathematics and physics, using them for engineering applications calls for a good understanding of the connections between scales, resolutions, and frequencies.

The term *scale* has a similar connotation to its usage in a geographical map. A map drawn to a large scale has fewer details than a map of the same region drawn to a smaller scale. Analogously, wavelet representations of signals at larger scales carry fewer details of the signal features than that at smaller scales. Continuing on the analogy, the wavelet transform of a signal at large scales has low or poor resolution of signal changes in time-domain, with the benefit of good localization of signal features in the low-frequency band. Relatively, wavelets at lower scales offer the reverse trade-off in abeyance with the bandwidth–duration principle.

Note that the term *small* or *large* scale is relative to the scale of the mother wavelet, for which $s=1$. Practical aspects are discussed in a later section.

4.1.2 Filtering perspective

The wavelet transform in Eq. (3.20) can be rewritten in a convolution form

$$Wx(u,s) = x * \bar{\psi}_s(\tau) \quad \text{where } \psi_s(t) = \frac{1}{\sqrt{s}} \psi^* \left(\frac{-t}{s}\right) \quad [3.24]$$

Thus, CWT is equivalent to filtering the signal by a filter whose IR is $\bar{\psi}_s(t)$ (Mallat, 1999).

Figure 3.5 illustrates the filtering nature of wavelet transforms using the Morlet wavelet (only the real part of the wavelet) and its Fourier transform. Scaling the mother wavelet automatically shifts the center frequency of the analyzing wavelet and also changes its T–F localization (spread) in accordance with the duration–bandwidth principle.

It is clear that the scale and Fourier frequency share an inverse relationship. The exact relationship between the scale and frequency depends on the center frequency ω_c of the mother wave. Torrence and Compo (1998) derive the relationship between the scale and Fourier period (wavelength) λ for different wavelets. For the Morlet wavelet with center frequency $\omega_0 = 6$, $s \approx (\omega_0 + \sqrt{2+\omega_0^2})/(4 * pi) \approx \lambda$.

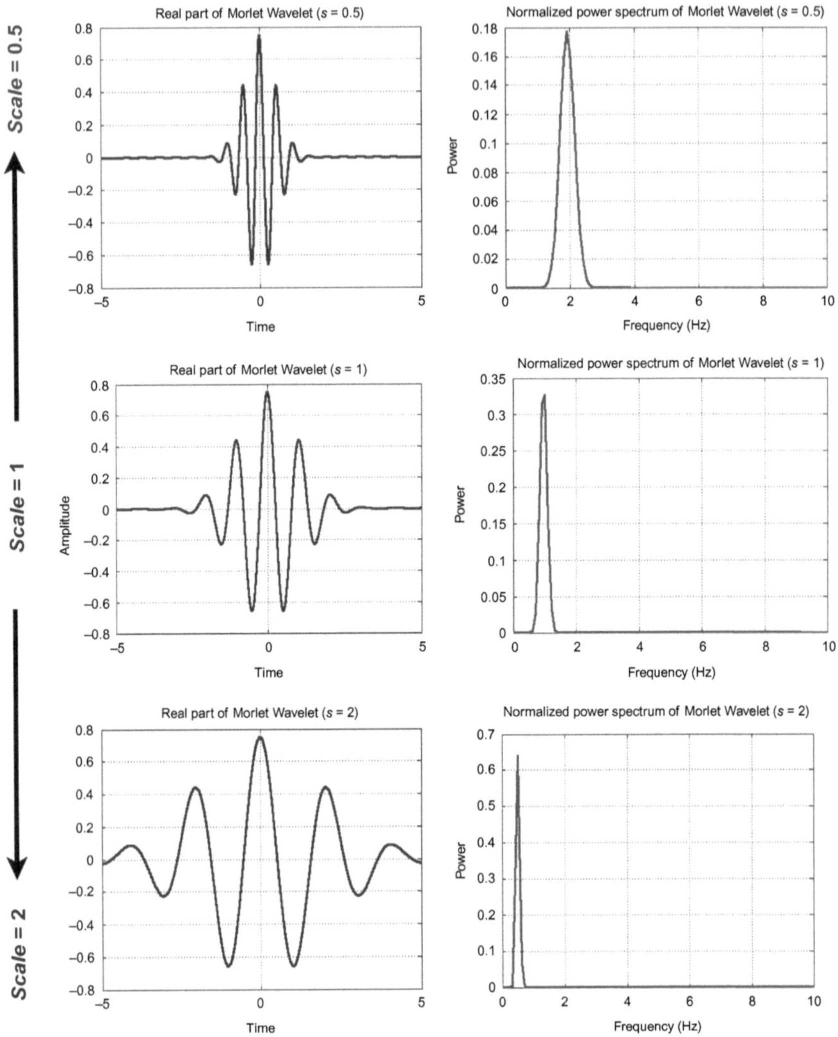

Figure 3.5 Scales $s > 1$ generate low (band)-pass filter wavelets while scales $s < 1$ generate high (band)-pass filter wavelets. Figures are shown for Morlet wavelet with center frequency $\omega_0 = 6$. (For color version of this figure, the reader is referred to the online version of this chapter.)

Qualitatively speaking, by setting $s = 1$ (the mother wave) as the reference point, the projections onto wavelet basis at scales $1 \leq s < \infty$ can be treated as approximations (low-frequency) and the projections at scales $0 < s < 1$ as the details corresponding to the approximation.

The filtering perspective leads us to the notion of *scaling functions*, as discussed later.

4.1.3 Scaling function

The family of wavelet bases generated by spanning the scaling factor from $0 < s < \infty$ and the translation parameter τ spans the entire $\mathbb{R}^2$ space (Mallat, 1999). As seen above, they also generate filters that span the entire frequency domain $\infty < \omega < 0$. For implementation purposes, the Fourier frequency axis is divided into $0 \leq \omega \leq \omega_0$ (low-frequency) and $\omega_0 < \omega < \infty$ (high-frequency), where ω_0 is the center frequency of the wavelet. Next, the infinite set of band-pass filters (wavelets) corresponding to $1 \leq s < \infty$ are replaced by a single low-pass filter while retaining the filters as is for $0 < s < 1$. From a functional analysis viewpoint, the $\mathbb{R}^2$ space is divided into an approximation space plus a detail space.

To determine the single low-pass filter that replaces the band-pass filters corresponding to $1 \leq s < \infty$, a *scaling function* is introduced such that (Mallat, 1999)

$$|\hat{\phi}(\omega)|^2 = \int_1^{\infty} |\hat{\psi}(s\omega)|^2 \frac{ds}{s} = \int_{\omega}^{\infty} \frac{|\hat{\psi}(\xi)|^2}{\xi} d\xi \qquad [3.25]$$

From the admissibility condition (3.22),

$$\lim_{\omega \to 0} |\hat{\phi}(\omega)|^2 = C_{\psi} \qquad [3.26]$$

it is clear that the scaling function $\phi(t)$ is a *low-pass* filter and only exists if Eq. (3.22) is satisfied, that is, if C_{ψ} exists. The phase of this low-pass filter can be chosen arbitrarily.

Equation (3.25) can be understood as follows. The aggregate of all details at "high" scales constitute an approximation. The aggregate of all the remaining details at lower scales constitute the details not contained in that approximation.

The scaling function $\phi(t)$ can also be scaled and translated like the wavelet function to generate a family of child scaling functions. The approximation *coefficients* of $x(t)$ at any scale are the projection coefficients of $x(t)$ onto the scaling function $\phi(t)$ at that scale

$$Lx(\tau, s) = \left\langle x(t), \phi_{\tau,s}(t) \right\rangle = x * \bar{\phi}_s(\tau) \qquad [3.27]$$

where L is the approximation operator. Generalizing the foregoing ideas by relaxing the reference point $s = 1$ that partitions the scale space, the inverse wavelet transform (IWT) in Eq. (3.21) can be broken up into two parts: an approximation at scale $s = s_0$ and all the details at scales $s < s_0$,

$$x(t) = \underbrace{\frac{1}{C_\psi s_0} Lx(.,s) * \phi_{s_0}(t)}_{\text{Approximation at scale } s_0} + \underbrace{\frac{1}{C_\psi}\int_0^{s_0} Wx(.,s) * \psi_s(t)\frac{\mathrm{d}s}{s^2}}_{\text{Details missed out by the approximation}} \qquad [3.28]$$

Equation (3.28) lays the foundations for multiresolution approximations (Mallat, 1989a) where the approximation term at each scale is further decomposed into a coarser approximation (higher scale) and detail in a nested manner.

4.1.4 Scalogram

The energy preservation equation (Eq. 3.23) provides the definition of *scalogram*, which has the same role as a spectrogram (of STFT) or a periodogram (of the FT). It provides the energy density in the time-scale or in the T–F plane. The scalogram in the T–F plane is defined as

$$P\left(\tau, \omega = \frac{\varsigma}{s}\right) = \left|Wx\left(\tau, \frac{\varsigma}{\omega}\right)\right|^2 \quad (\text{time} - \text{frequency plane}) \qquad [3.29]$$

where ζ is the conversion factor from $1/s$ to frequency. Based on the discussion in Section 4.1.2, ζ largely depends on the center frequency.

A normalized scalogram $\frac{1}{s}P(\tau,\omega) = \frac{\omega}{\zeta}P(\tau,\omega)$ (Addison, 2002; Mallat, 1999) facilitates better comparison of energy densities at two different scales by taking into account the differences in widths of the wavelets at two different scales. Figure 3.6 illustrates the benefit of using the normalized scalogram for the case of a mix of two sine waves with periods $T_{\mathrm{p}1} = 5$ and $T_{\mathrm{p}2} = 20$. The unnormalized version presents an incorrect picture of the relative energy of the two components.

The scalogram is the central tool in CWT applications to T–F analysis. Section 6 reviews the underlying ideas and applications to control and modeling of systems.

It is appropriate to compare the performance of scalogram with that of spectrogram for the example used to generate Fig. 3.2. The scalogram for the example is shown in Fig. 3.7. Unlike in STFT where a special effort is required to select the appropriate window length, the wavelets at lower scales are naturally suited to detecting time-localized features in a signal while those at higher scales are naturally suited for frequency-localized features.

In Figs. 3.6 and 3.7, a cone like profile is observed. This is called the *cone of influence* (COI) (Mallat, 1999; Torrence and Compo, 1998). The COI

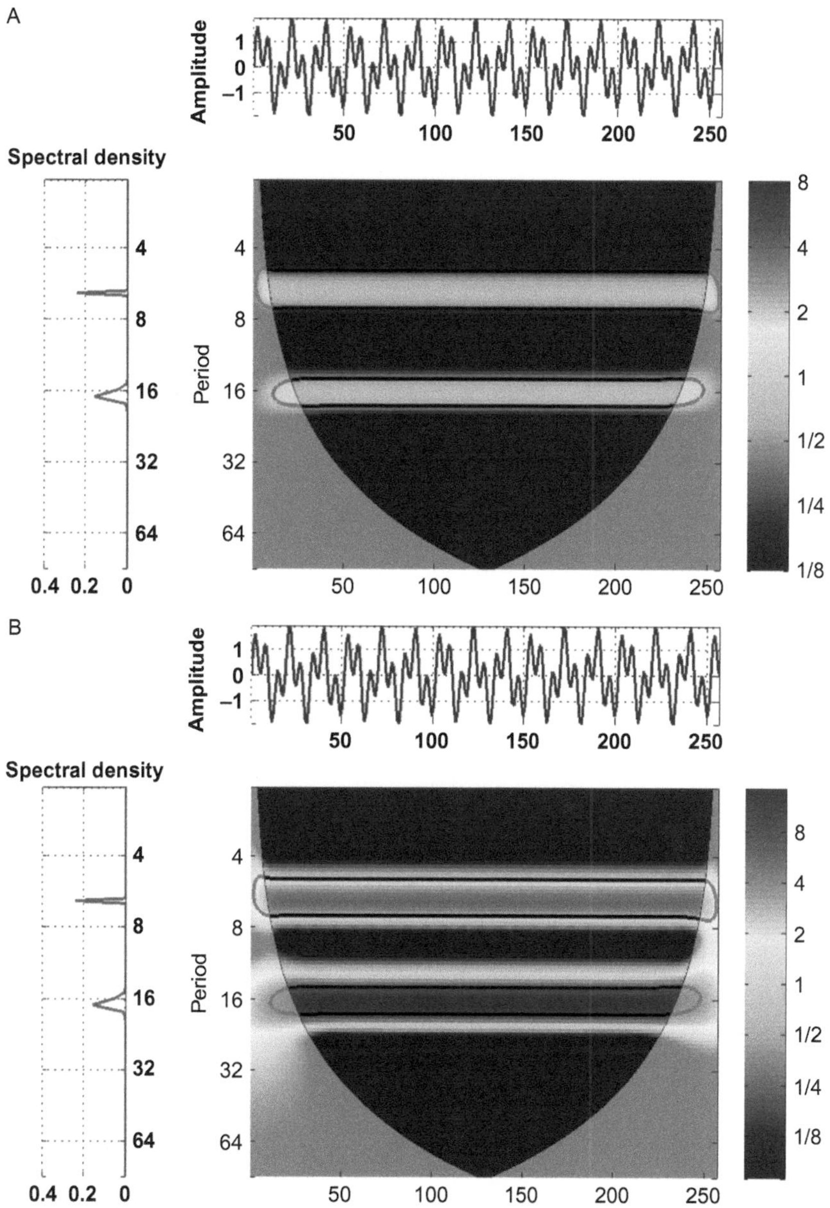

Figure 3.6 Normalization facilitates a correct comparison of energy densities at two different scales. (A) Normalized scalogram and (B) unnormalized scalogram. (For color version of this figure, the reader is referred to the online version of this chapter.)

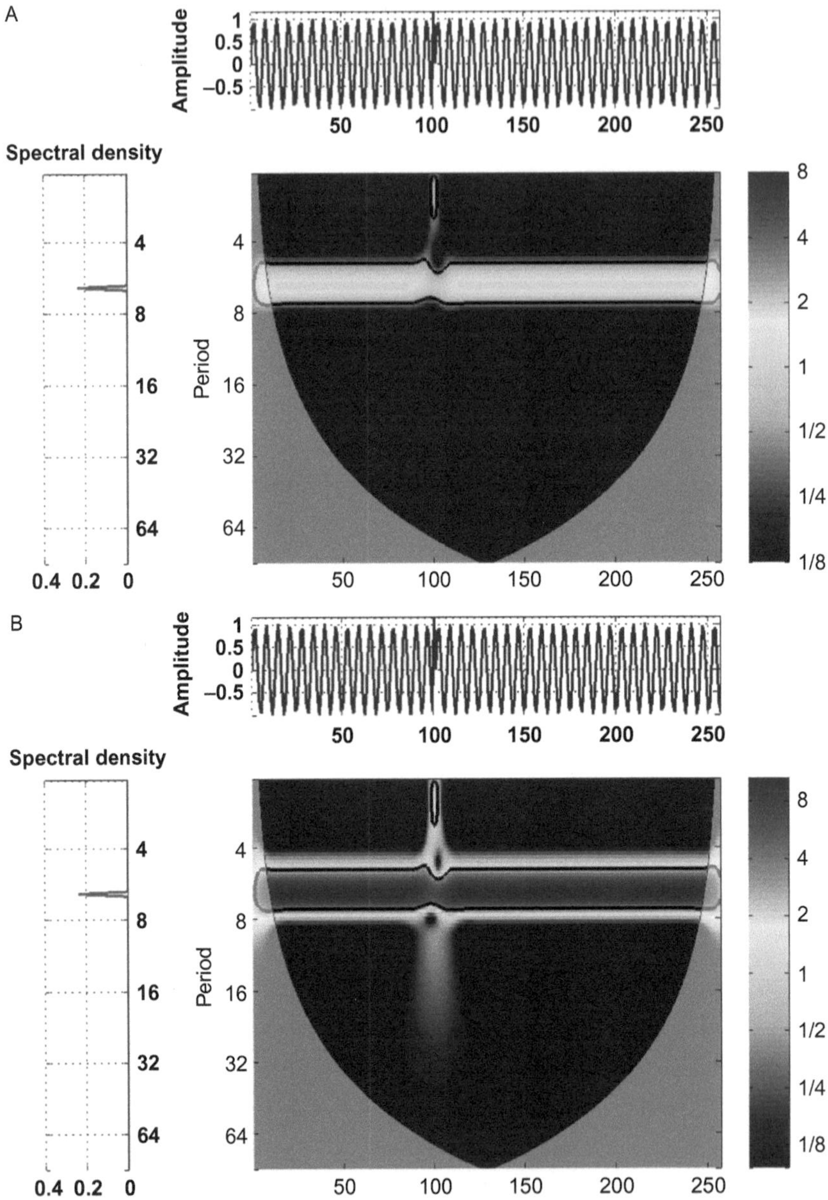

Figure 3.7 Scalogram detects the presence of impulse located at $k=100$ very well. (A) Normalized scalogram and (B) unnormalized scalogram. (For color version of this figure, the reader is referred to the online version of this chapter.)

arises because of the finite length data and the border effects of wavelets at every scale. The effect depends on the scale since the length of the wavelet that is outside the edges of the signal is proportional to the length of the scale. A useful interpretation of COI is that it is the region beyond which the edge effects are negligible. A formal treatment of this topic can be found in Mallat (1999).

4.1.5 Choice of wavelets

Several wavelet families exist depending on the choice of the mother wave, each catering to a specific need. Recall that the choice of basis is largely driven by the application, that is, the signal features that are of interest. Wavelet families can be primarily categorized into four classes:

1. *(Bi)orthogonal wavelets*: These are useful for filtering and multiresolution analysis. They produce a compact representation of the signal.
2. *Nonorthogonal wavelets*: These wavelets are useful for time-series analysis and result in a highly redundant representation.
3. *Real wavelets*: Real-valued wavelets are used in detecting peaks or discontinuities or measuring regularities of a signal.
4. *Complex wavelets*: This class of wavelets is useful for T–F (phase and amplitude of the oscillatory components) analysis of signals.

Figure 3.8 depicts six of the popularly used wavelet basis functions. Two of these wavelet functions, namely, Mexican hat and Morlet wavelets, do not possess scaling functions counterparts since they do not satisfy the admissibility condition (22), that is, C_ψ does not exist for these wavelets. Wavelets can also be characterized by three properties, namely, (i) compact support, (ii) vanishing moments, and (iii) symmetry.

A closed-form (explicit) expression for wavelets does not necessarily always exist. Where a closed-form does not exist, the IR coefficients of the associated filter are specified.

The Morlet wavelet is a complex wavelet characterized by

$$\psi(t) = \pi^{-1/4}\left(e^{j\omega_0 t} - e^{-\omega_0^2/2}\right)e^{-t^2/2} \approx \pi^{-1/4} e^{j\omega_0 t} e^{-t^2/2} \quad [3.30]$$

$$\Rightarrow \hat{\psi}(\omega) = \pi^{1/4}\sqrt{2}e^{-(\omega-\omega_0)^2/2} \quad [3.31]$$

where ω_0 is the center frequency of the wavelet. It is widely used in the T–F analysis of signals. The center frequency governs the frequency of the signal component that is being analyzed. It does not have a compact support but has a fast decay.

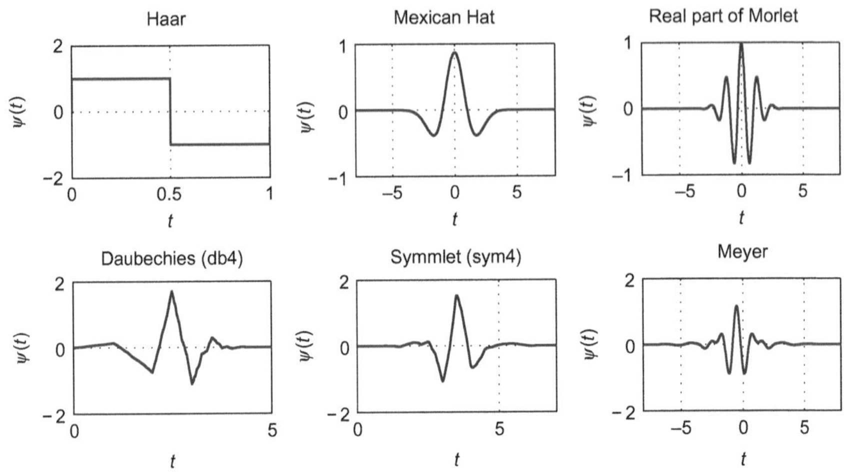

Figure 3.8 Different wavelet functions possessing different properties. (For color version of this figure, the reader is referred to the online version of this chapter.)

On the other hand, Daubechies wavelets are a class of real, continuous, orthogonal wavelets characterized by the IR coefficients of the associated high-pass filter. The length of the filter influences the *vanishing moments*' property of the wavelet,

$$\int_{-\infty}^{+\infty} t^n \psi(t)\mathrm{d}t = 0 \quad \text{for } 0 \leq n < p \qquad [3.32]$$

which is related to the degree of polynomial that a wavelet can exactly explain. This property is useful in capturing the regularity (smoothness) of $f(t)$. It can be proved that if $f(t)$ is regular and $\psi(t)$ has enough vanishing moments then the wavelet coefficients $\langle f, \psi_{j,k} \rangle$ are small at fine scales.

Conventionally, Daubechies wavelets are denoted by *dbp*, where p is the vanishing moments of the wavelet. Although asymmetric and not possessing linear phase, they possess the minimum support for a given vanishing moments. The Haar wavelet can be treated as a special case of Daubechies wavelets with a single vanishing moment but is discontinuous in nature.

See Mallat (1999) for an extensive treatment of the different types of wavelets, their properties, and uses. There is no single wavelet suited for all applications. The choice is largely governed by the end-use requirements. An extensive discussion on a suitable choice of mother wavelet from information-theoretic considerations is contained in Gao and Yan (2010, chapter 10).

4.1.6 Computation of CWT

The CWT of a signal $x(t)$ can be computed efficiently using the Fourier Transform route (Addison, 2002; Gao and Yan, 2010; Torrence and Compo, 1998).

Recalling Eq. (3.20) and using the fact that convolution transforms to product in the Fourier domain, the Fourier transform of CWT is computed first, followed by the Inverse FT.

$$\mathcal{F}[W_x(\tau,s)] = \sqrt{s}X(\omega)\hat{\psi}^*(s\omega) \qquad [3.33]$$

$$\Rightarrow W_x(\tau,s) = \frac{1}{2\pi}\int_{-\infty}^{\infty}\sqrt{s}X(\omega)\hat{\psi}^*(s\omega)e^{j\omega\tau}d\omega \qquad [3.34]$$

In practice, a discrete version of the above is implemented by evaluating CWT over a user-specified grid of scales and translations. It usually results in a highly redundant representation of the signal in the time-scale space.

For a comprehensive and insightful understanding of the use of CWT in data analysis, the reader is directed to the short and insightful guide by Torrence and Compo (1998). The guide elucidates various aspects relevant to the implementation and interpretation of CWT in practice.

4.2. Discrete wavelet transform

The discrete wavelet transform is the CWT evaluated at specific scales and translations, $s=2^j, j\in\mathcal{Z}$ and $\tau=m2^j$, $m\in\mathcal{Z}$.

$$Wf(m,j) = \int_{-\infty}^{+\infty} f(t)\psi^*_{m2^j,2^j}(t)dt \qquad [3.35]$$

where

$$\psi_{m2^j,2^j}(t) = \frac{1}{2^{j/2}}\psi\left(\frac{t-m2^j}{2^j}\right) \qquad [3.36]$$

DWT provides a compact (minimal) representation, whereas CWT offers a highly redundant representation. By restricting the scales to octaves (powers of 2) and translations proportional to the length of the wavelet at each scale, a family of orthogonal wavelets is generated. When the restrictions on translations alone are relaxed, a *dyadic* wavelet transform is generated, which once again presents a complete and stable, but a redundant representation. The frame theory offers a powerful framework for characterizing completeness, stability, and redundancy of a general basis representation in inner product spaces (Daubechies, 1992; Daubechies et al., 1986; Duffin and Schaeffer, 1952).

The dyadic discretization of scales and translations not only impart orthogonality to DWT but also render a very important attribute, which is that of multiresolution approximations. The MRA stems from the fact that the *approximations constructed at two successive dyadic scales are related through a scaling relation*. Consequently, one can progressively construct and recover a set of embedded approximations of a signal at different resolutions (scales of approximation). This is very attractive for computer vision, image processing, and modeling multiscale systems.

The use of DWT gained significant momentum with the discovery of connections between orthogonal wavelet transforms and multirate filter banks formalized in the works of Mallat (1989b) and conditions on perfect reconstruction filters (Smith and Barnwell, 1986; Vaidyanathan, 1987; Vetterli, 1986). It was further propelled by the arrival of Daubechies' wavelets (Daubechies, 1988) which were the first orthogonal, continuous wavelets with compact support to be designed. Mallat's work (Mallat, 1989b) laid down the platform for several engineering applications and forms the basis of most practical implementations of wavelet transforms today. The main contributions were the formalization of MRA and a fast pyramidal algorithm for wavelet transform through a series of low-pass and high-pass filtering plus downsampling operations.

A brief review of MRA and the associated theory follows.

4.3. Multiresolution approximations

Given that the focus is on approximations, a good starting point for presenting MRA is the projections onto the space spanned by scaling functions.

The scaling functions at two different scales $s=2^j$ and $s=2^{j+1}$ have widths proportional to 2^j and 2^{j+1}, respectively. The resolving ability of the scaling function is inversely proportional to its width. Therefore, the scaling function at a higher scale will have a lower resolving ability than that at a lower scale. The multiresolution approximation of signals is a family of approximations of a signal generated at different resolutions, but with an important requirement. *The approximation at a lower resolution should be embedded in the approximation at a higher resolution*. In other words, the basis space spanned by the translates of $\phi(t/2^{j+1})$ should be contained in the space spanned by the translates of $\phi(t/2^j)$.

Transferring the above requirement to the basis functions for the respective spaces, we embark upon the popular two-scale relation (or the dilation relation, see Strang and Nguyen, 1996),

$$\frac{1}{\sqrt{2}}\phi\left(2^{-(j+1)}t\right) = \sum_{n=-\infty}^{+\infty} h[n]\phi(2^{-j}t - n) \quad [3.37]$$

The right-hand side (RHS) has a convolution form. Therefore, the coefficients $\{h[n]\}_{n\in\mathbb{Z}}$ can be thought of as the IR coefficients of a filter that produces a coarser approximation from a given approximation.

From Section 2 and Appendix A, approximation of $x(t)$ at a level j is its orthogonal projection onto the subspace spanned by $\{\phi(2^{-j}t-n)\}_{n\in\mathbb{Z}}$, which is denoted by V_j. Then the detail at that level is contained in the subspace W_j. At a coarser level $j+1$, the approximation lives in the subspace V_{j+1} with a corresponding detail space W_{j+1}. MRA implies V_{j+1}, $W_{j+1} \subset V_j$. Specifically,

$$V_j = V_{j+1} \oplus W_{j+1}, \; j \in \mathbb{Z} \quad [3.38]$$

$$\mathcal{P}_{V_j}x = \mathcal{P}_{V_{j+1}}x + \mathcal{P}_{W_{j+1}}x. \quad [3.39]$$

Thus, W_{j+1} contains all the details to move from level $j+1$ to a finer level j. It is also the orthogonal complement of V_{j+1} in V_j.

A formalization of these ideas due to Mallat and Meyer can be found in many standard wavelet texts (see Mallat, 1999; Jaffard et al., 2001).

A function $\phi(t)$ should satisfy certain conditions in order for it to generate an MRA. A necessary requirement is that the translates of $\phi(t)$ should be linearly independent and produce a stable representation, not necessarily energy-preserving and orthogonal. Such a basis is called *Riesz* basis (Strang and Nguyen, 1996).

The central result is that the requirements on $\phi(t)$ can be expressed as conditions on the filter coefficients $\{h[n]\}$ in the dilation equation (Eq. 3.37) (Mallat, 1999). Some excerpts are given below.

4.3.1 Filters and MRA

Where an *orthogonal* basis is desired, the conditions on the filter are (Mallat, 1999; Meyer, 1992)

$$|\hat{h}(\omega)|^2 + |\hat{h}(\omega+\pi)|^2 = 2; \;\; \hat{h}(0) = \sqrt{2} \;\; \forall \omega \in \mathbb{R} \quad [3.40]$$

Such a filter $\{h[n]\}$ is known as the *conjugate mirror filter* (Smith and Barnwell, 1986; Vetterli, 1986). Notice that $\hat{h}(\pi) = 0$.

Practically, the raw measurements are at the finest time resolution and assumed to represent level 0 approximation coefficients (note that sampling is also a projection operation). A level 1 approximation is obtained by projecting it onto $\phi(t/2)$ (level $j = 1$). The corresponding details are generated by projections onto the wavelet function $\psi(t/2)$. This is a key step in MRA.

By the property of the MRA, the space spanned by $\psi(2^{-(j+1)t})$ (coarser scale) should be contained in the space spanned by translates of $\phi(2^{-j}t)$ (finer scale). Hence,

$$\frac{1}{\sqrt{2}}\psi\left(2^{-(j+1)}t\right) = \sum_{n=-\infty}^{+\infty} g(n)\phi\left(2^{-j}t - n\right) \qquad [3.41]$$

Interestingly, once again $\{g[n]\}_{n\in\mathbb{Z}}$ can be thought of as the IR coefficients of a filter that produces the details corresponding to the approximation generated by $\{h[n]\}_{n\in\mathbb{Z}}$.

Corresponding to the conditions of Eq. (3.40), for any $\left\{\psi_{n,j}(t)\right\}_{n,j\in\mathbb{Z}}$ to generate an orthonormal basis while satisfying Eq. (3.41), the filter $\{g[n]\}$ should satisfy (Mallat, 1999; Meyer, 1992)

$$\hat{g}(\omega) = \mathrm{e}^{-i\omega}\hat{h}^*(\omega + \pi) \Rightarrow g[n] = (-1)^{1-n}h[1-n] \qquad [3.42]$$

Thus, the filters $h[n]$ and $g[n]$ are tied together. Moreover, observe

$$\hat{h}(0) = \sqrt{2} \Rightarrow \hat{g}(0) = 0 \qquad [3.43]$$

giving them the characteristics of a low- and high-pass filter, respectively.

From a filtering viewpoint, the relation (Eq. 3.42) between low- and high-pass filters of the wavelet transforms and the fact that different frequency components of the signal can be extracted in a recursive manner sets them apart from the traditional scheme of filtering.

Interestingly, all other important requirements, namely, compact support, vanishing moments, and regularity, can be translated to conditions on the filters $h[n]$ and $g[n]$ (Mallat, 1999). For example, compact support of $\phi(t)$ requires $h[n]$ also to have compact support and over the same interval. Thus, the design of scaling and wavelet functions essentially condenses to design of associated filters.

4.3.2 Reconstruction

Quite often one may be interested in reconstructing the signal as is or its approximation depending on the applications. In estimation, this is a routine step. Decompose the measurement up to a desired level (scale). If the details at that scale and finer scales are attributed to noise, then recover only that portion of the measurement corresponding to the approximation. For these and related purposes, reconstruction filters $\widetilde{h}[n]$ and $\widetilde{g}[n]$ are required.

Perfect reconstruction requires that the filters $\widetilde{h}[n]$ and $\widetilde{g}[n]$ satisfy (Vaidyanathan, 1987)

$$g[n] = (-1)^{1-n}\widetilde{h}[1-n] \quad \widetilde{g}[n] = (-1)^{1-n}h[1-n] \qquad [3.44]$$

With an orthonormal basis (conjugate mirror filters), it can be shown that the reconstruction filters are identical to the decomposition filters, that is, $\widetilde{h}[n] = h[n]$ and $\widetilde{g}[n] = g[n]$. Daubechies filters are examples of this class.

4.3.3 Biorthogonal wavelets

If the decomposition and reconstruction filters are different from each other, then the basis for MRA is nonorthogonal. A special and useful class of filters emerges when we require

$$\left\langle \widetilde{h}[k], h[k-2n] \right\rangle = \delta[n] \quad \langle \widetilde{g}[k], g[k-2n] \rangle = \delta[n] \qquad [3.45]$$

$$\left\langle \widetilde{h}[k], g[k-2n] \right\rangle = 0 \quad \langle \widetilde{g}[k], h[k-2n] \rangle = 0 \qquad [3.46]$$

These filters are known as *biorthogonal* filters (see Mallat, 1999 for a detailed exposition). In terms of approximation detail spaces, W_j is no longer orthogonal to V_j but is orthogonal to $\widetilde{V}_j$. Similarly, $\widetilde{W}_j$ is only orthogonal to V_j. A classic example of bi-orthogonal wavelets is the one that is derived from the *B*-spline scaling function. Later in this work, we use biorthogonal wavelets for modeling. Some discussion on spline wavelets is therefore warranted.

Polynomial splines of degree $l \geq 0$ spanning a space V_j are set of functions that are $l-1$ times differentiable and equal to a polynomial of degree l in the interval $[m2^j, (m+1)2^j]$. A Riesz basis of polynomial splines of degree l is constructed by starting with a *box spline* $\mathbf{1}_{[0,1]}$ and convolving with itself l times. The resulting scaling function is then a spline of degree l having a Fourier transform

$$\hat{\phi}(\omega) = e^{-j\frac{\varepsilon\omega}{2}}\left(\frac{\sin\left(\frac{\omega}{2}\right)}{\frac{\omega}{2}}\right)^{l+1} \qquad [3.47]$$

whose associated low-pass filter is specified in the Fourier domain

$$\hat{h}(\omega) = \sqrt{2}e^{-j\frac{\varepsilon\omega}{2}}\left(\cos\frac{\omega}{2}\right)^{l+1} \qquad [3.48]$$

The corresponding time-domain filter coefficients $h[n]$ and the reconstruction filter coefficients $\tilde{h}[n]$ are available in the literature (see Mallat, 1999, chapter 7). The orthogonal spline functions were independently introduced by Battle (1987) and Lemarie (1988); however, the basis does not have a compact support. On the other hand, the semiorthogonal (only orthogonality across scales) *B*-spline wavelets of Chui and Wang (1992) and Unser et al. (1996) have compact support, but either by the analysis or by the synthesis basis. However, the biorthogonal splines due to Cohen et al. (1992) possess compact support. They are one of the most popular classes of spline wavelets.

Spline biorthogonal wavelets are popularly known as reverse biorthogonal (RBIO) wavelets and are designated as *rbio* $p{\cdot}\tilde{p}$ or *spline* $p{\cdot}\tilde{p}$. Figure 3.9 graphs the scaling and wavelet functions corresponding to the synthesis and reconstruction RBIO filters. These wavelets sacrifice the orthogonality within $(\phi(t), \psi(t))$ and $\tilde{\phi}(t)$, $\tilde{\psi}(t)$ but offer a number of attractive features such as best approximation ability among all the wavelets of an order l, explicit expressions in time- and frequency domains, compact

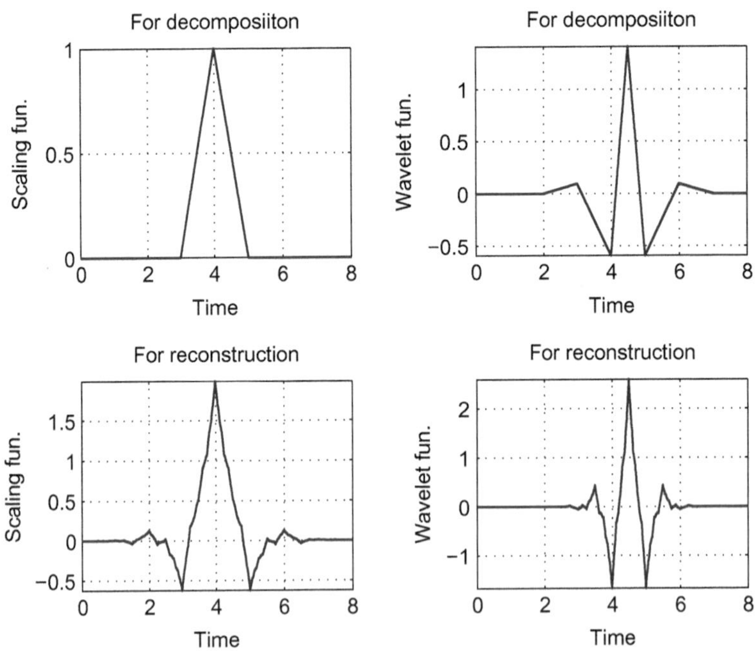

Figure 3.9 Spline biorthogonal scaling functions and wavelets of vanishing moments $p=2$ and $\tilde{p}=4$ for the decomposition and reconstruction wavelets, respectively. (For color version of this figure, the reader is referred to the online version of this chapter.)

support with optimal T–F localization, etc. The reader is directed to Unser (1997) for a scholarly exegesis of this topic.

4.4. Computation of DWT and MRA

The DWT and hence the MRA are computed by means of a fast algorithm due to Mallat (1989b). A review of this algorithm follows.

First, recall that the projections are

$$\mathcal{P}_{V_j}x = \sum_{n=-\infty}^{\infty} \left\langle x, \phi_{n,j} \right\rangle \phi_{n,j} \quad \mathcal{P}_{W_j}x = \sum_{n=-\infty}^{\infty} \left\langle x, \psi_{n,j} \right\rangle \psi_{n,j} \qquad [3.49]$$

which are characterized by the approximation and detail *coefficients* as

$$a_j[n] \triangleq \left\langle x, \phi_{n,j} \right\rangle \quad d_j[n] \triangleq \left\langle x, \psi_{n,j} \right\rangle \qquad [3.50]$$

The coefficients $a_j[n]$ and $d_j[n]$ carry approximation and detail information of x at the scale 2^j, respectively.

By virtue of MRA, the approximation and detail coefficients at a higher scale can be computed from the approximation coefficients at a finer (lower) scale,

$$\begin{aligned} a_{j+1}[k] &= \sum_{n=-\infty}^{\infty} h[n-2k]a_j[n] = \left(a_j * \bar{h}\right)[2k]; \quad d_{j+1}[k] \\ &= \sum_{n=-\infty}^{\infty} g[n-2k]a_j[n] = \left(a_j * \bar{g}\right)[2k] \end{aligned} \qquad [3.51]$$

where $\bar{h}[n] = h[-n]$ and $\bar{g}[n] = g[-n]$.

The unusual convolution in the above equation is implemented as a combination of regular convolution and downsampling operations.

Reconstruction of coefficients from a given approximation and detail coefficients involves convolution of upsampled coefficients and the reconstruction filters.

The fast algorithm due to Mallat (1989b) is essentially based on the above ideas. It offers a computationally efficient means of computing the decompositions and reconstructions at different scales.

Decomposition

1. Assume the discrete-time signal $x[n]$ to be the approximation coefficients of a continuous-time signal $x(t)$ at $j=0$, that is, set $x[n] \equiv a_0[n]$ $\Rightarrow x(t) = \sum_{n=-\infty}^{\infty} a_0[n]\phi(t-n) \in V_0$. This is the finest resolution for the

signal. At this level, further details are unavailable. Therefore, $d_0[n] = 0\ \forall n$

2. Compute the coarse approximation and the details upto a desired level J by recursively implementing equations in Eq. (3.51). Figure 3.10 illustrates the fast pyramidal algorithm for orthogonal wavelet decomposition.

The convolution in Eq. (3.51) is implemented in two steps consisting of filtering plus downsampling (decimation) by a factor of 2. The downsampling essentially (i) removes redundancy in the filtered sequences $\tilde{a}_j$ and $\tilde{d}_j$ and (ii) accounts for translations of $\phi(2^{-j}t)$ in steps of 2^j at a given scale. The hallmark of the transform is that downsampling does not cause any loss of information. The following example inspired by Murtagh (1998) serves to illustrate these ideas. *Example:* Consider a signal sequence $\{x[1], x[2], \ldots, x[N]\}$ whose MRA we wish to construct.

Choose Haar filters Haar (1910) $h[n] = \left[(1/\sqrt{2})(1/\sqrt{2})\right]$ and $g[n] = \left[(1/\sqrt{2})(-1/\sqrt{2})\right]$ respectively.

The filtered data sequence is

$$\tilde{a}_1 = \frac{x[1]+x[2]}{\sqrt{2}}, \frac{x[2]+x[3]}{\sqrt{2}}, \frac{x[3]+x[4]}{\sqrt{2}}\ldots, \text{ and}$$

$$\tilde{d}_1 = \frac{x[1]-x[2]}{\sqrt{2}}, \frac{x[2]-x[1]}{\sqrt{2}}, \frac{x[3]-x[2]}{\sqrt{2}}\ldots,$$

Next, downsample by a factor of 2 to obtain the *scaling* and *detail* coefficients at level $j=1$,

$$a_1 = \frac{x[1]+x[2]}{\sqrt{2}}, \frac{x[3]+x[4]}{\sqrt{2}}, \frac{x[5]+x[6]}{\sqrt{2}}\ldots, \text{ and}$$

$$d_1 = \frac{x[1]-x[2]}{\sqrt{2}}, \frac{x[3]-x[4]}{\sqrt{2}}, \frac{x[5]-x[6]}{\sqrt{2}}\ldots$$

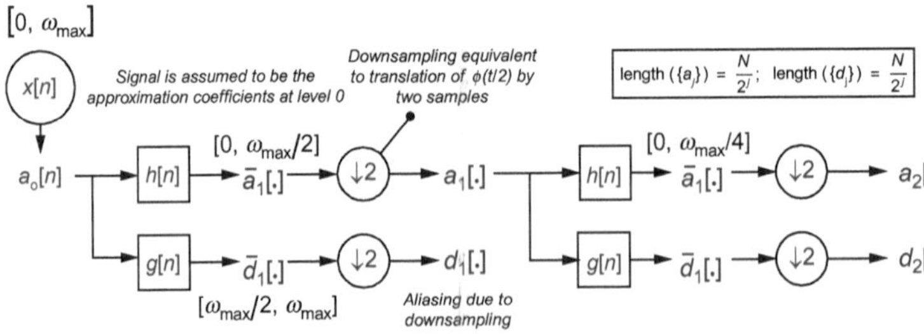

Figure 3.10 Fast pyramidal algorithm for orthogonal wavelet decomposition. (For color version of this figure, the reader is referred to the online version of this chapter.)

The original sequence can be still reconstructed with this downsampled sequence:

$$x[1]=\frac{a_1(1)+d_1(1)}{\sqrt{2}},\ x[2]=\frac{a_1(1)-d_2(1)}{\sqrt{2}}\cdots,$$

The multiscale approximation can be constructed until a desired level J. The coarsest approximation coefficients are obtained at $J_{\max}=\log_2 N$.

An important remark is in place here. The time-interval spanned by $\{a_1\}$ and $\{d_1\}$ is identical to the time-interval of $x[n]$. Moreover, the cardinality is preserved, meaning the total number of coefficients in a_j and $\{d_j\}_{j=1}^{M}$ equals N, the total number of samples in $x[k]$. However, the time resolution falls off by a factor of two with increase in level. The time stamps corresponding to the coefficients depend on the phase of the wavelet and scaling functions. See Percival and Walden (2000) for an in-depth treatment of this subject.

Reconstruction

Recovery of the signal proceeds in the opposite direction. Figure 3.11 depicts the algorithm used for reconstruction. Using the approximation coefficients at level $j=L$ and detail coefficients at levels $j=L, L-1, \ldots, 1$, one can perfectly recover the signal at the finest scale by a series of upsampling and filtering (with the reconstruction filters) operations. The upsampling (insertion of zeroes) is necessary to cancel the frequency folding (aliasing) created during downsampling in the decomposition stage Mallat (1999).

In the preceding example, the reconstruction expressions for $x[1]$ and $x[2]$ yield the same as upsampling a_1 and d_1 by a factor of 2 (insert zero between samples) followed by convolution with $\tilde{h}=\left[(1/\sqrt{2})\,(1/\sqrt{2})\right]$ and $\tilde{g}=\left[(1/\sqrt{2})\,(-1/\sqrt{2})\right]$.

In general, the component of the measurement corresponding to a desired scale of approximation or detail can be reconstructed separately. This is achieved by ignoring all the other coefficients (approximation and/or detail) at other scales in the reconstruction. Figure 3.12A and B illustrates

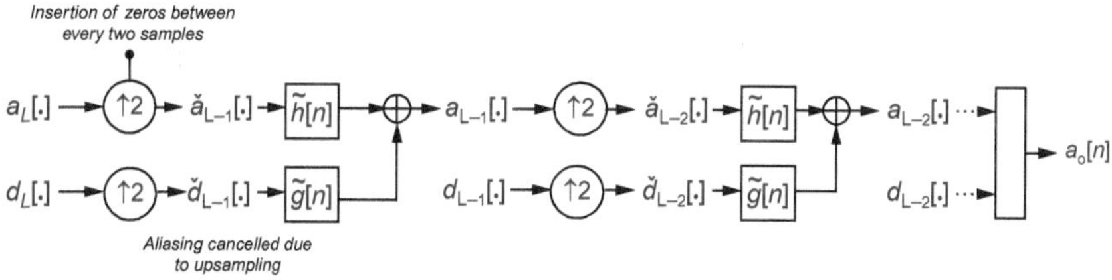

Figure 3.11 Fast algorithm for reconstruction from decomposed sequences. (For color version of this figure, the reader is referred to the online version of this chapter.)

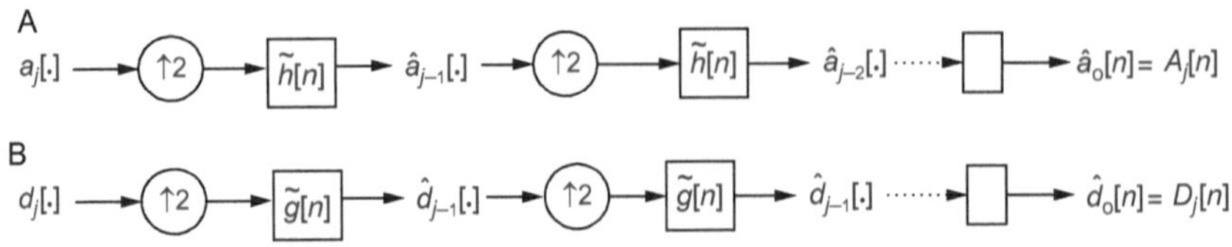

Figure 3.12 DWT facilitates separate reconstruction of low- and high-frequency components at each scale. (A) Reconstruction of components in the low-frequency band (approximations) of the *j*th level and (B) reconstruction of components in the high-frequency band (details) of the *j*th level. (For color version of this figure, the reader is referred to the online version of this chapter.)

these ideas. The reconstructed low- and high-frequency sequences corresponding to the *j*th level are denoted by $\mathcal{A}_j$ and $\mathcal{D}_|$, respectively.

By the linearity of the transform and virtue of MRA,

$$\mathbf{x} = \mathcal{A}_1 + \mathcal{D}_1 = \mathcal{A}_2 + \mathcal{D}_2 + \mathcal{D}_1 = \mathcal{A}_M + \sum_{j=M}^{1} \mathcal{D}_j \qquad [3.52]$$

In terms of expansion coefficients,

$$x(t) = \sum_m \alpha_{j_0,m}\phi_{j_0,m}(t) + \sum_{j \geq j_0}\sum_m \beta_{j_0,m}\psi_{j,m}(t) \qquad [3.53]$$

It is instructive to compare Eq. (3.53) with the CWT version in Eq. (3.28) by setting $s_0 = 2^M$. Thus, once again the information in $x(t)$ is reordered in terms of the coefficients $\alpha(.)$ and $\beta(.)$. The filtering perspective explains the tiling of the T–F plane by the DWT as shown in Fig. 3.3. The ability to break up a signal into approximation and details at a desired set of scales and reconstruct the signal in parts or whole empowers wavelet transforms with the ability to segregate components of a multiscale, compress, denoise and estimate signals in an optimal manner.

A simple example using a synthetic signal (Department of Statistics, 2000; Mallat, 1999) is shown to illustrate the foregoing ideas.

Example

A piecewise-regular polynomial (Mallat, 1999) is taken up for illustration. The approximation and detail coefficients from a three-level Haar wavelet decomposition are shown in Fig. 3.13A. The top panel shows the signal under analysis. Observe that the discontinuities reflect in the highest frequency band. The trend is captured in $256/2^3 = 32$ approximation coefficients at the third level $\mathbf{a}_3$. These 32 coefficients contain 88.4% of the signal's energy. In Fig. 3.13B, the components of the signal corresponding

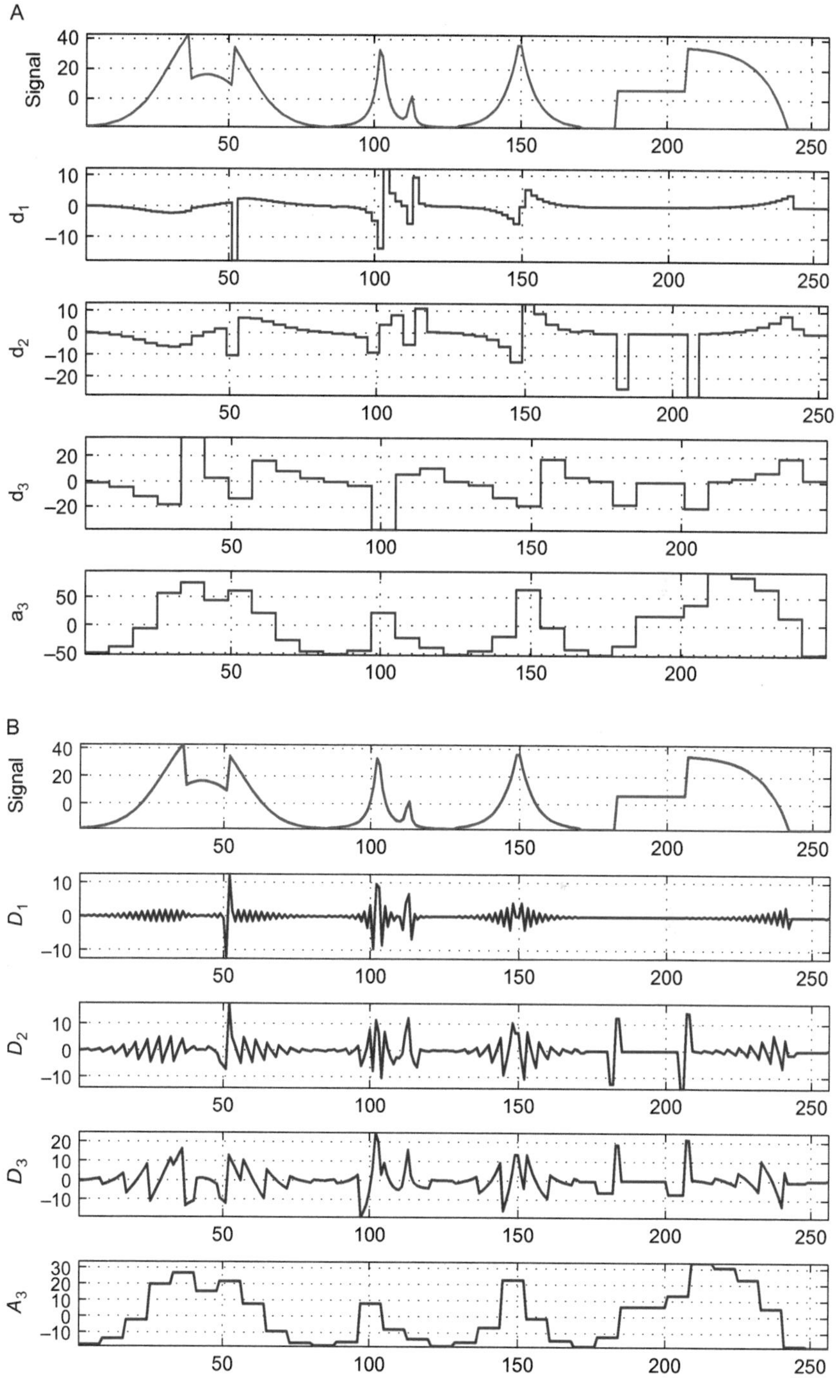

Figure 3.13 Wavelet decomposition and MRA of a piecewise regular polynomial (Mallat). (A) Three-level Haar decomposition, (B) reconstructed components, and

(Continued)

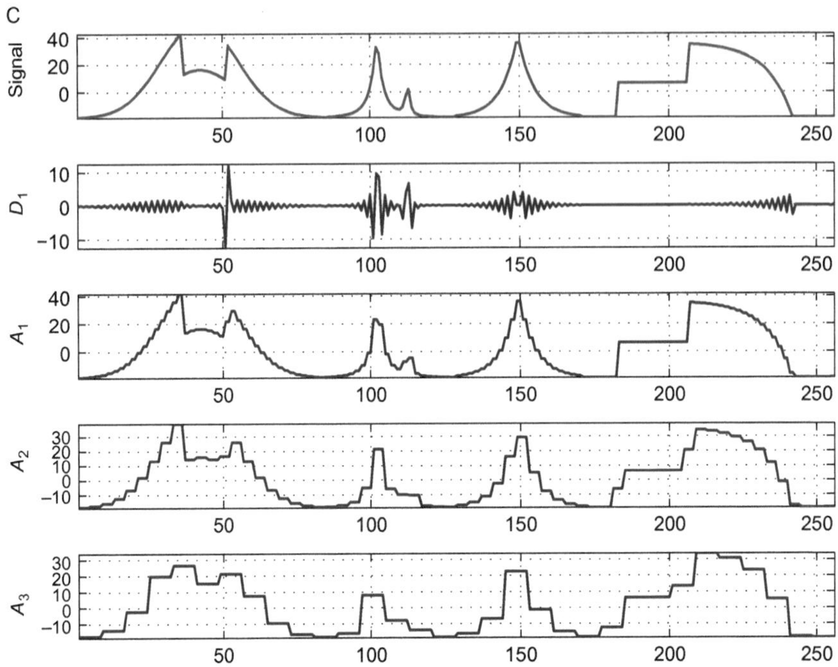

Figure 3.13—Cont'd (C) multiresolution approximation. (For color version of this figure, the reader is referred to the online version of this chapter.)

to the approximation and detail coefficients of Fig. 3.13A are reconstructed. When an MRA is desired, the approximations at each scale are reconstructed by ignoring the details at that scale and finer (lower) scales. Figure 3.13C shows the MRA of the example signal starting from the coarsest (third level) moving up to the first level. The zeroth level approximation is the signal itself. Observe how features are progressively added as one moves from coarser to finer approximations. The details left out by the finest approximation A_1 are also shown. As in Fig. 3.13A, the top panel shows the signal.

4.4.1 Features of wavelet coefficients

The wavelet coefficients, particularly the DWT coefficients, possess a number of useful and interesting properties:

1. Correlation functions in wavelet domain decay faster than those of the original measurement in time (Tewfik and Kim, 1992). In the context of data analysis, this feature is largely possessed only by the wavelet coefficients since the approximation coefficients (detail) contain the

deterministic signal characteristics, which usually belong to the low-frequency bands. This property is exploited by modeling and monitoring techniques that work with wavelet domain representations of signals.

2. The coefficients at a scale j contain the energy contributions due to changes in signal at that scale owing to the energy decomposition of the signal (Parseval's result for DWT)

$$\|\mathbf{x}\|_2^2 = \|\mathbf{a}_J\|_2^2 + \sum_{j=1}^{J} \|\mathbf{d}_j\|_2^2 \quad [3.54]$$

Further, this is true of the reconstructed sequences as well,

$$\|\mathbf{x}\|_2^2 = \|\mathcal{A}_J\|_2^2 + \sum_{j=1}^{J} \|\mathcal{D}_j\|_2^2 \quad [3.55]$$

The energy decomposition is used in time-series analysis of multiscale signals (see Percival and Walden, 2000) and in other applications (Addison, 2002; Rafiee et al., 2011; Unser and Aldroubi, 1996).

3. DWT provides a sparse representation of signals (measurements). Most of the information in the measurement is contained in a few scaling function coefficients. Most of the noise content of is spread among the detail coefficients. The compression algorithms based on DWT exploit this property quite effectively by only storing the approximation coefficients and the thresholded detail coefficients (see Chau et al., 2004; Vetterli, 2001).

4. Discontinuities and nonlinearities in signals are highlighted by the high-frequency band coefficients (finer scales), while the regular parts of the signal are highlighted by the approximation coefficients. The ability to detect discontinuities largely depends on the choice of wavelets. A Haar wavelet is appropriate for this purpose. A generalization of these ideas is the use of modulus maxima of the wavelet coefficients for singularity detection. See Mallat (1999) for an illustration of these ideas.

4.5. Other variants of wavelet transforms

With the CWT and DWT as the base, a number of variants of wavelet transforms have come to the forefront, a majority of them being based on DWT owing to its tremendous potential in a diverse set of fields. Popular among these are the wavelet packet transform (WPT) and the maximal overlap DWT (or the shift-invariant DWT). These variants extend the applicability of DWT to signal estimation and pattern recognition by incorporating specific features into the DWT. Once again, the modifications can be summed up as a different ways of tiling the T–F plane.

The presentation on WPT and maximal overlap DWT (MODWT) below is strictly to provide the reader with the breadth of the subject. Space constraints do not permit a tutorial style exposition of the topics. The reader is referred to Mallat (1999), Percival and Walden (2000), and Gao and Yan (2010) for a gradual and in-depth development of these variants.

4.5.1 Wavelet Packet Transform

The WPT is a straightforward extension of the DWT to arrive at a more flexible/adaptive signal representation. The difference is essentially that unlike in DWT the detail space W_j is also split into an approximation and detail space along with V_j. Consequently, the frequency axis is divided into smaller intervals. The signal decompositions are therefore in packets of frequency intervals and hence the name. In addition, the analyst can choose to split the approximation and details at select scales. Alternatively, a full decomposition can be performed on the signal, following which only select frequency bands can be retained for reconstruction. These features impart enormous flexibility in signal representation and the way the T–F plane is tiled.

Figure 3.14 is illustrative of the underlying ideas in WPT.

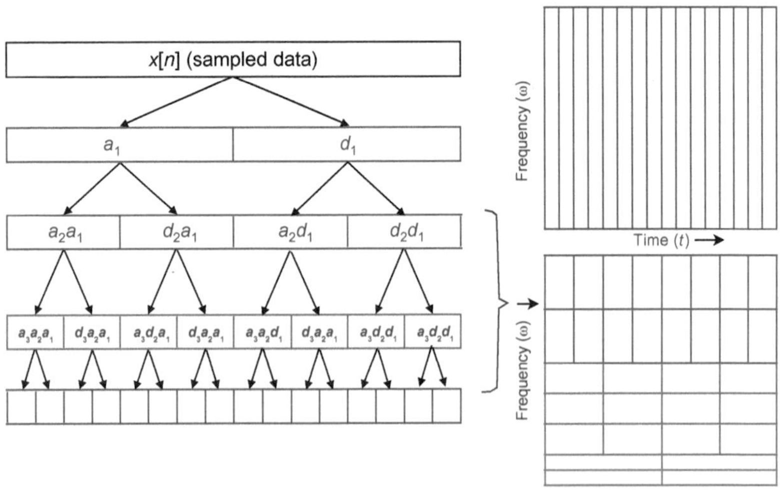

Figure 3.14 WPT tiles the frequency plane in a flexible manner and facilitates the choice of frequency packets for signal representation. (For color version of this figure, the reader is referred to the online version of this chapter.)

The WPT essentially decomposes the signal into components localized in different frequency subbands by projecting both the approximation and detail coefficients onto coarser spaces. The basis for each of these subbands is known as wavelet packets. As in the case of wavelet transforms, downsampling of high-frequency subbands at any stage results in frequency folding. Therefore, a frequency reordering of the decomposed components is necessary, which can be related to Gray coding of binary strings (Gray, 1953). The tiling achieved by the WPT in the T–F plane is shown on the right bottom of Fig. 3.14. In general, an arbitrary tiling, still bounded by the bandwidth–duration principle, can be achieved. It is instructive to compare the tiling with that of the DWT.

The purpose of using a WPT is select the most appropriate subbands for an "optimal" signal representation. A "best" basis search algorithm based on some cost criterion (e.g., entropy) is deployed for this purpose. In Fig. 3.14, the highlighted subbands on the left could be the best set of frequency bands for a given signal. Splitting the signal into finer subbands other than the selected bands would cause the cost function to increase and is hence optimal. The division of the T–F plane by the choice of the basis in these frequency bands is shown on the right side of Fig. 3.14. Thus, the WPT localizes the energy in Heisenberg boxes (tiles in T–F plane) in an adaptive manner in contrast to DWT which always decomposes the energy into predetermined boxes.

WPT finds wide applications in signal estimation, image analysis, and feature extraction. For an extensive treatment of the underlying ideas and a collection of related applications, the reader may refer to Addison (2002), Gao and Yan (2010), and Mallat (1999).

4.5.2 Maximal overlap DWT

A major shortcoming of the DWT is that it is not shift-invariant, meaning a *shift of a signal feature in time does not produce the same time-shift in the wavelet coefficients* (see Percival and Walden, 2000 for a nice illustrated example). This is not surprising since the time axis sampling is not dense. To introduce the shift-invariant property, the wavelet windows are only translated by one sampling interval while still retaining the dyadic discretization of the scale parameter. Thus, the windows at any scale have a maximal overlap giving the transform its name, MODWT. It is also known by other names: translation (shift)-invariant DWT, dyadic wavelet transform, and undecimated DWT. The basis functions responsible for this transform are, expectedly, not orthonormal.

This variant of the transforms finds extensive use in analysis of time series and modeling. Implementation of MODWT is performed using the same algorithm as for DWT with the omission of the downsampling (and upsampling) steps (Mallat, 1999; Percival and Walden, 2000).

4.6. Fixed versus adaptive basis

A remarkable distinction can be observed between the FT, the STFT, the WVD, and the wavelet transforms (including its variants). Section 2 emphasized the fact that the transforms are essentially projections onto basis functions. In choosing the basis functions, two routes are possible: (i) a *fixed basis* set, where the user has a complete knowledge of the basis functions and (ii) an *adaptive basis*, where the user derives the basis set from the signal. The Fourier transform and its short-time variants deploy a fixed basis, whereas the WVD can be viewed as the transform with a basis derived from the signal.

Wavelet transforms belong to the class of fixed basis. Nevertheless, in the literature, one often associates wavelet transforms (particularly the WPT) with the "adaptive basis" class of methods. This can be misleading unless the term "adaptive" is properly understood. The adaptivity is largely *a posteriori* effect, that is, the user can choose the basis to be retained for reconstruction or representation with the help of certain optimization criteria. Nevertheless, this does not make it truly adaptive since the shape of the basis functions in the selected frequency bands is fixed known *a priori*. Thus, wavelet transforms are at best semiadaptive.

In passing, an important cautionary remark is in order. Although wavelet transforms find a vast number of applications due to their versatility, it is not a panacea to all the problems of multiscale analysis. It is essential to understand its limitations. Wavelets are not the ideal tools when signals contain short-lived, low-frequency components or whose energy densities vary along a polynomial in the T–F plane (e.g., chirps). While the WPT offers some improvements in this regard, the (pseudo) WVD offers a much better T–F localization of the energy density. Figure 3.15 illustrates this point in case. The signal consists of three amplitude-modulated Gaussian atoms in series. The pseudosmoothed WVD gives the best picture of the energy density when compared to the one obtained from STFT and CWT. It may be noted that WVDs may be a poor choice for signals with a different set of features. Moreover, it may also be recalled that WVDs are not ideally suited to filtering applications.

We close this section with a reference to a recently evolved method for multiscale signal analysis known as the Hilbert Huang transform (HHT)

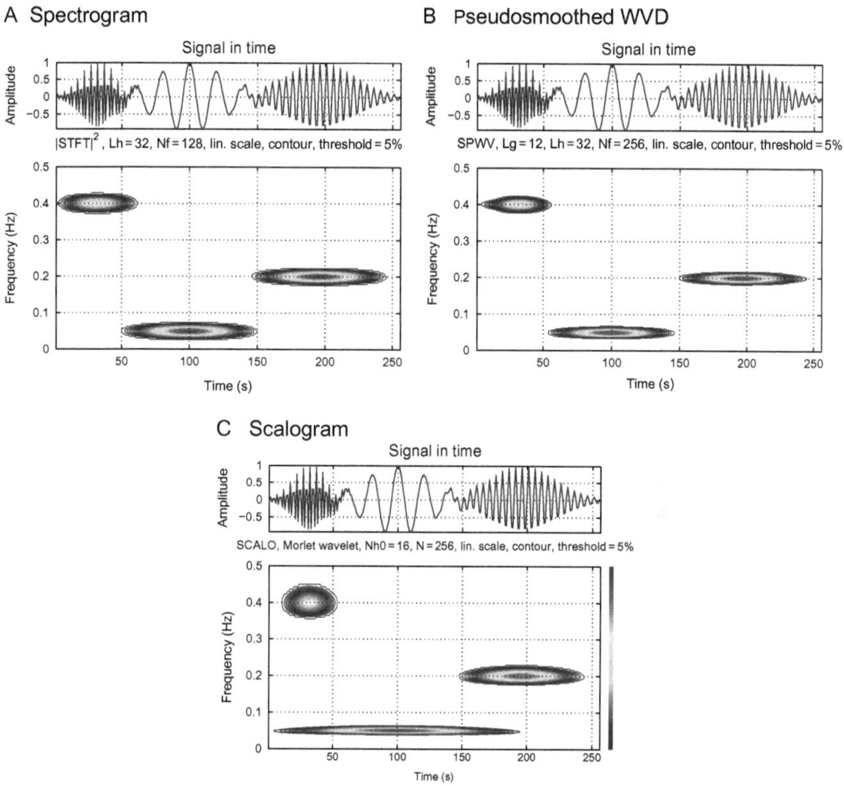

Figure 3.15 Synthetic example: Wavelets may not be the best tool for every application. (A) Spectrogram, (B) pseudosmoothed WVD, and (C) scalogram. (For color version of this figure, the reader is referred to the online version of this chapter.)

based on the idea of *empirical mode decomposition* (EMD) (Huang et al., 1998). The HHT, also like wavelet transform, breaks up the signal into components that are *analytic*, with the help of EMD, and subsequently performs a Hilbert transform of the components. The HHT belongs to the adaptive basis class of methods and in principle has the potential to be superior to WT. However, it is computationally more expensive and lacks the transparency of the WT.

4.7. Applications of wavelet transforms

Wavelet transforms lend themselves to an enormous number and a diverse set of applications such as filtering, T–F analysis, multiscale approximations, signal compression, solutions to differential equations, modeling of non-stationary stochastic processes, etc. Table 3.1 gives a short glimpse of the multifaceted potential of WT.

Table 3.1 Areas and applications of wavelet transforms

Geophysics	Atmospheric and ocean processes, climatic data
Engineering	Fault detection and diagnosis, process identification and control, nonstationary systems, multiscale analysis
DSP	Time–frequency analysis, image and speech processing, filtering/denoising, data compression
Econometrics	Financial time-series analysis, statistical treatment of wavelet-based measures
Mathematics	Fractals, multiresolution approximations (MRA), wavelet-based nonparametric regression, solutions to differential equations
Medicine	Health monitoring (ECG, EEG, neuroelectric waveforms), medical imaging, analysis of DNA sequences
Chemistry	Flow injection analysis, chromatography, IR, NMR, UV spectroscopy data, quantum chemistry
Astronomy	MRA of satellite images, solar wind analysis

Any attempt to review the entire breadth of engineering applications of these transforms in a single article would be futile. The discussion is restricted to the applications of wavelets to control loop performance monitoring and modeling. Several control and modeling applications of WTs deploy wavelets for signal estimation either as a preprocessor or as an intermediate step. It is therefore appropriate to begin with a brief review of the same. Through the brief review, we take the opportunity to draw the reader's attention to a relatively less-used method of signal estimation, known as *consistent estimation*.

5. WAVELETS FOR ESTIMATION

5.1. Classical wavelet estimation

Signal estimation is concerned with the problem of recovering the signal from its measurements, which are corrupted with noise and disturbances. The term *denoising* is synonymously used with estimation in the wavelet literature. Estimation of signals (or parameters/states) is one of the most crucial exercises in data analysis. In Section 2, the idea of estimating a signal by a thresholding of the Fourier coefficients was discussed. Wavelet denoising essentially works on the same principle and is also the basis of the pioneering works of Donoho (1995) and Donoho et al. (1995). The wavelet denoising method is a highly well-established technique for signal estimation with attractive features. The method produces near optimal nonlinear estimate of the signal (Mallat, 1999).

A primitive estimate of the signal is obtained by completely discarding coefficients at finer scales that predominantly carry effects of noise followed by a reconstruction of the retained coefficients. Clearly, this can be detrimental in many situations since the finer scales may carry information on abrupt changes in the process, sudden sensor failures, edge information, etc. (Bakshi, 1999). Therefore, the idea of thresholding is employed. All state-of-the-art denoising algorithms consist of three steps: (i) decomposition, (ii) thresholding, and (iii) reconstruction of the thresholded coefficients. The core step is thresholding, for which a variety of methods are available differing in the way the threshold is determined and applied. Four threshold estimation algorithms are popular, namely, (i) universal (Donoho et al., 1995), (ii) minimax (Donoho and Johnstone, 1994), (iii) Stein's unbiased risk estimator (SURE) (Stein, 1981), and (iv) minimum description length (MDL) (Stein, 1981). An intuitive way of applying the threshold is to set all coefficients below the threshold to zero (hard thresholding). Additionally, one can shrink the magnitude of the retained coefficients by the threshold value (soft thresholding) (Donoho, 1995). Five variants of these ideas are prominent, viz., (i) global, (ii) level dependent, (iii) data dependent (iv) cycle-spin (translation invariant), and (v) WPT-based thresholding methods. The thresholding approach to denoising can be nicely shown to be the solution to the classical $\|.\|_2$-norm minimization estimation problem with a $\|.\|_1$-norm penalization of the wavelet coefficients (see Percival and Walden, 2000 for a pedagogical development).

The success of any denoising algorithm depends on how closely the signal and noise characteristics agree with the assumptions of a particular method.

Figure 3.16A shows the result of denoising a (mean-centered) level measurement from a simulated industrial process (Tangirala et al., 2005). A soft thresholding with global threshold method assuming scaled white noise is employed for this purpose. In Fig. 3.16B, the measurement of weigh feeder controller in an industrial process is cleaned using a Symmlet-3 with a four-level universal soft-thresholding denoising method, assuming white-noise measurement error. Observe that the important features of the signal are preserved in the cleaned signal.

An excellent comparative study of the different wavelet denoising techniques combining 22 different wavelet choices, 4 threshold estimation methods, and 4 different threshold application methods applied to synthetic and chemometric signals is reported in Cai and Harrington (1998). The

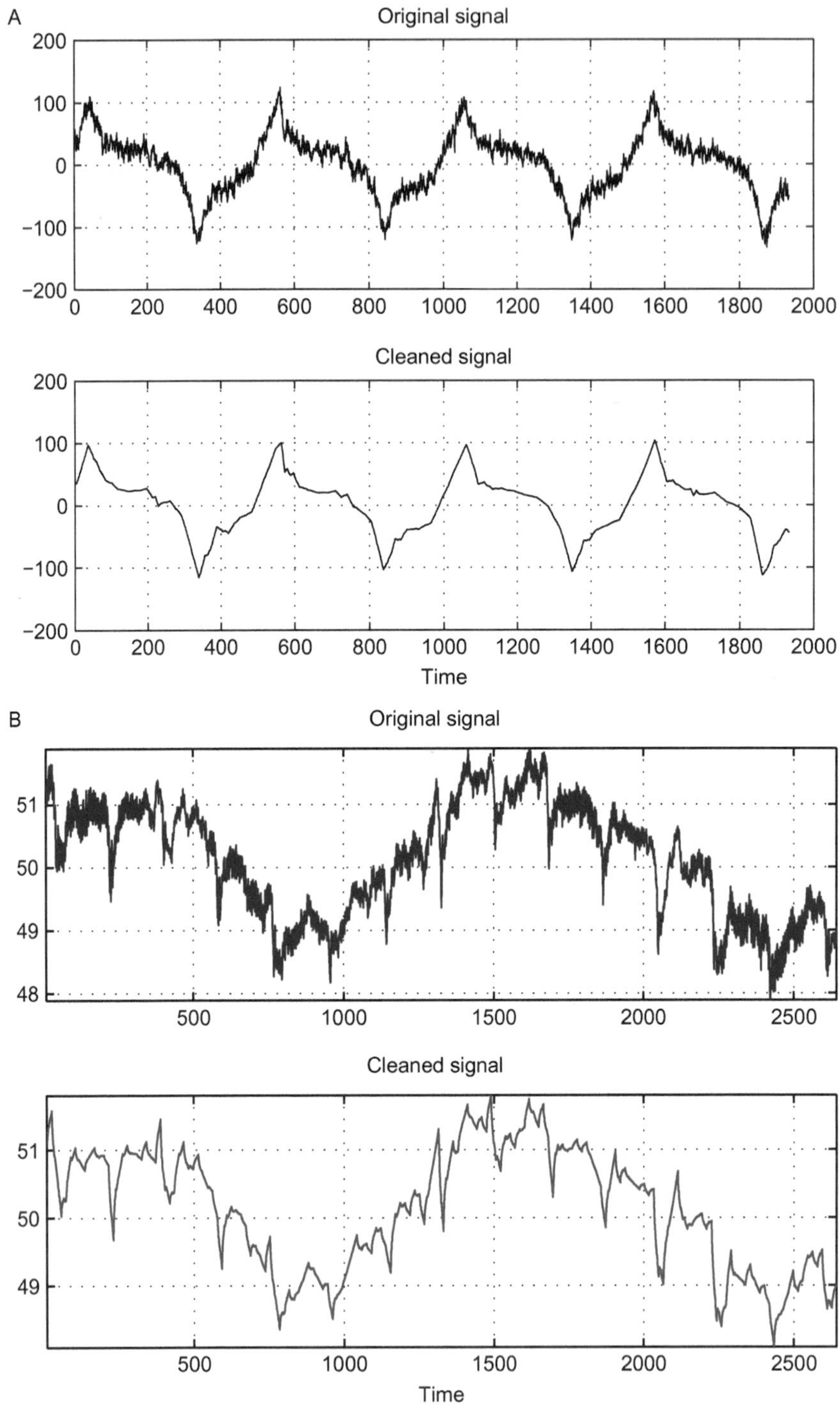

Figure 3.16 Original measurement and denoised signals. (A) Level deviations in a simulated industrial process and (B) weigh feeder controller output in an industrial process. (For color version of this figure, the reader is referred to the online version of this chapter.)

study advocates the use of translation invariant method of applying the threshold that is determined by a MDL algorithm. In another relatively less exhaustive study, Rosas-Orea et al. (2005) conduct a comparison of three denoising algorithms using wavelets on synthetic and real data. The conclusions differ not only with respect to the study by Cai and Harrington (1998) but also across data. For synthetic data, their study concludes the choice of rigorous SURE algorithm with a hard threshold as best suited, whereas best performance for real data is given by a universal soft-thresholding algorithm with a *db*5 wavelet.

Majority of the denoising applications in chemical engineering have been for outlier detection and noise removal (Bakshi and Nounou, 2000; Nounou and Bakshi, 1999). Nounou and Bakshi (1999) combine wavelet thresholding techniques with multiscale median filtering for online filtering of random and gross errors as well as data rectification. Denoising principles are also used in compression applications because of the strong similarity in the governing ideas. Most control and modeling applications use wavelet denoising as a preliminary or an intermediate step.

5.2. Consistent estimation

Consistent estimation of a signal is an alternative and perhaps an advanced way of signal estimation introduced in the works of Cvetkovic and Vetterli (1995) and Thao and Vetterli (1994). The ideas underlying consistent estimation though already existed in the works by Mallat and Zhong (1992). A signal $\hat{x}[k]$ is a *consistent estimate* of the original signal $x[k]$ if it possesses the same representation (in a specified domain) as the original signal. The term *representation* is defined as follows.

Definition

A signal representation in transform (or measurement) domain is an ordered collection of significant signal values in that domain (obtained by a nonlinear operation such as maxima detection or thresholding).

In other words, the signal representation is a pair consisting of an index (in the domain) and the associated signal value with the indices arranged in ascending order. For example, the thresholded wavelet coefficients of a measurement (of a noisy signal) form a representation of the signal in wavelet domain because thresholding removes noise coefficients. Another example is the representation of the signal using the zero-crossing of its wavelet transform (Mallat, 1991).

Consistent estimation differs from classical estimation in that it explicitly forces the estimate to possess the same features as the signal in the *representation* domain. Furthermore, consistent estimation is carried out using the undecimated or the dyadic wavelet transform, MODWT, in contrast to the DWT that is used in traditional denoising methods. Finally, spline biorthogonal wavelets are normally recommended (Mallat and Zhong, 1992) for obtaining the consistent estimate, whereas the denoising methods admit any orthogonal wavelet basis.

The implementation consists of an alternate projection algorithm that switches back and forth between the time and wavelet domain to converge to a solution (Mallat and Zhong, 1992). In fact, the classical (wavelet denoising) wavelet estimation is only the first step of the iterative algorithm. The need for the switching between time and wavelet domains is that the thresholded coefficients, or, in general, a sequence of numbers (functions) $\{g_j(\cdot)\}_{j\in\mathcal{Z}}$ need not be *a priori* the wavelet transform of a signal (function) $f(.)$ (Mallat, 1991). In other words, it is not necessary that any sequence is the wavelet transform of a function, that is, it necessarily satisfies $(g_j(\cdot))_{j\in\mathcal{Z}} = \mathcal{W}f$. The following steps outline the alternate projection algorithm:

1. Perform dyadic wavelet transform of the signal $y[k]$. Call it $Y_w(.)$
2. Threshold the wavelet coefficients to obtain $\widetilde{Y}_w(\cdot)$, which is sparse. Store the indices corresponding to significant coefficients.
3. Operate $\mathcal{W}\mathcal{W}^{-1}$ on $\widetilde{Y}_w(\cdot)$, where $\mathcal{W}$ is the wavelet transform operator. Call this $\bar{Y}_w(\cdot)$.
4. Force the significant coefficients of $\bar{Y}_w(\cdot)$ to match the significant coefficients of $Y_w(.)$ at the stored indices.
5. Repeat steps 3 and 4 until convergence.

Convergence of the above algorithm is proved in Mallat and Zhong (1992). The solution obtained is optimal in the least squares sense.

The idea of consistent estimation is illustrated in Fig. 3.17 with an application to signal denoising. The top panel of Fig. 3.17A shows a synthetic noisy signal marked as measurement, obtained by adding a colored noise of signal-to-noise ratio (SNR) 30 dB to the original signal. The consistent estimate of the signal is shown in the bottom panel along with original signal. The estimate is obtained by a reconstruction of the thresholded wavelet projections of the noisy signal using the iterative alternate projection algorithm (Cvetkovic and Vetterli, 1995).

The wavelet projections at different scales of the original signal and the estimated signal (indicated by circles and solid lines, respectively) are shown in Fig. 3.17B. Theoretically, the projections of the reconstruction should

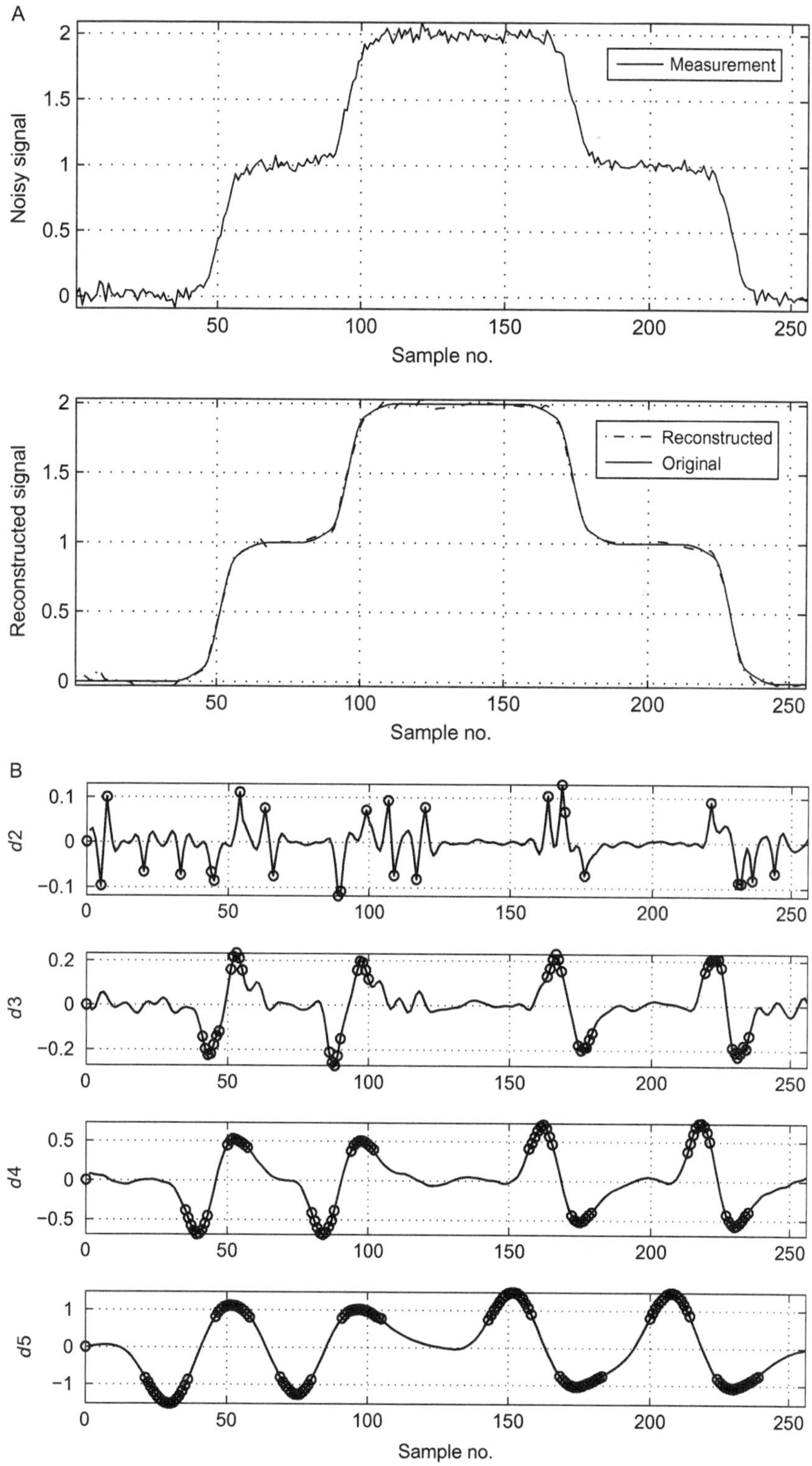

Figure 3.17 Example illustrating consistent estimation. (A) Noisy signal and its consistent estimate and (B) coefficients *a*4 and *d*4 to *d*2.

match that of the (original) signal at every point in the domain. However, since the reconstruction is obtained from a subset of projections (the original set is never known, unless the wavelet achieves perfect separation between the signal and noise), the matching occurs only at those select indices.

5.3. Signal compression

The ideas and techniques implemented in wavelet denoising carry forward to compression of data as well. Philosophical differences exist though. In signal estimation, the search for the threshold is to maximize the noise elimination while minimizing the damage to the signal. On the other hand, for compression, the optimal threshold is that which preserves as much as information as possible while still producing a compact representation of the signal. Moreover, compression avoids the step of reconstruction. Finally, signal estimation requires the wavelet to possess good separability property, whereas compression algorithms require the wavelet to be able to represent the measurement (or its predominant part) in as few coefficients as possible. Both requirements are related but not necessarily identical.

Signals compressed with wavelets can be combined with multivariate compression algorithms (e.g., principal component analysis). A combination of multivariate and univariate compression techniques for online compression of magnetic flux leakage signals in pipeline inspection is reported in a recent work by Kathirmani et al. (2012).

6. WAVELETS IN MODELING AND CONTROL

Identification, control, and monitoring of multiscale systems are usually more complicated and cumbersome than that of single-scale systems (Braatz et al., 2006; Ricardez-Sandoval, 2011). It is well known that a direct application of standard control methods, or even modern control methods to such multiscale system models, can lead to complicated or ill-conditioned controllers (high sensitivity), closed-loop instability, etc., (Christofides and Daoutidis, 1996; Kokotovic et al., 1986). To circumvent these problems, multiscale systems can be represented as a combination of models with fast and slow dynamics (Luse and Khalil, 1985). This decomposition is advantageous for the reason that the design criteria for the slow dynamics differ considerably from that of the fast dynamics. Moreover, the degree of accuracy with which slow dynamics are identified is quite higher than that with the fast dynamics. Singular perturbation theory has been found to offer a useful framework for modeling of multiscale systems (Khalil, 1987; Kokotovic et al., 1976, 1986; O'Reilly, 1980; Saksena et al., 1984).

With the emergence of wavelet transforms in the early 1980s, researchers found an excellent tool for effectively and elegantly describing multiscale, time-varying, and nonlinear systems in the T–F domain (e.g., see Bakshi, 1999; Motard and Joseph, 1994). Wavelets have since then been used in (i) theoretical and empirical modeling, (ii) formulating new control algorithms, (iii) monitoring multiscale systems, and (iv) online gross-error detection and filtering. A majority of these methods employ the discretized versions of the transform, in particular, the DWT. The undecimated and WPT also occupy an appreciable place. Techniques based on CWT are relatively scarce. In the discussions to follow, we confine ourselves to modeling and control applications. Specifically, the *focus is on empirical modeling (system identification) and control loop performance monitoring applications.*

The use of wavelet transforms in the applications of modeling and control can be subdivided into three classes, namely, (i) T–F analysis-based methods, (ii) methods that exploit the multiscale filtering ability of wavelets, and (iii) methods that employ wavelets as basis functions. We explore these applications in the following sections.

6.1. Wavelets as T–F (time-scale) transforms

6.1.1 Controller loop performance monitoring

The literature on the use of wavelet-based time-frequency representation methods for control and closed-loop performance monitoring (CLPM) is isolated. CLPM is concerned with (i) evaluating the performance of control loops and (ii) diagnosing the cause(s) of poor loop performance (Desborough and Miller, 2002; Jelali, 2005). The performance of control loops can be below par or degrade due to a combination of factors—poor controller tuning, oscillatory disturbances, actuator nonlinearities, sensor malfunctions, and model-plant mismatch (Choudhury et al., 2010; Desborough and Miller, 2002; Harris et al., 1999; Selvanathan and Tangirala, 2010). The assessment step is concerned with detecting the poor performance using a suitable benchmark, which is a challenge in itself (Jelali, 2005). Process delay is a vital piece of information necessary for assessment (Harris, 1989). Diagnosis is even more challenging because one needs to know the mapping between the sources of poor performance and the performance metrics. The literature on CLPM and, in particular, diagnosis is replete with ideas and applications (Jelali, 2005; Srinivasan and Tangirala, 2010; Thornhill and Horch, 2007). The general idea is to search for signatures or features in the manipulated and controlled variables of the poorly performing loops. For example, valve nonlinearities manifest as harmonics in the spectral signature of the output (Choudhury et al., 2005), whereas aggressive controller tuning can produce oscillations at a single frequency (Thornhill et al., 2003). Parametric approaches have also been proposed, wherein specific model structures for describing process and actuator characteristics are estimated (see, e.g., Srinivasan et al., 2005). Parameters of these identified models under some assumptions can reveal the source of poor performance.

Wavelets are applied in CLPM for detection of plant-wide oscillations, delay estimation, and diagnosis of poor loop performance. Oscillations in controlled outputs are clear indicators of poor loop performance, in general suboptimal plant performance and economic losses (Thornhill et al., 2003). Plant-wide oscillations are also cause for safety concerns. Causes for oscillations can be one or more among aggressive controller tuning, actuator nonlinearities, oscillatory disturbances, propagated effects, and model-plant mismatch (Jelali, 2005; Selvanathan and Tangirala, 2010; Thornhill et al., 2003). On the other hand, it is possible that these oscillations do not persist throughout but can be intermittent due to a combination of reasons. Figure 3.18A and B shows the scalograms of two different measurements of a refinery process (Tangirala et al., 2007; Thornhill et al., 2003). The scalogram of the downstream measurement reveals presence of persistent oscillations in that measurement, whereas the scalogram of the upstream measurement shows that oscillations are intermittent. The cause for the intermittent disappearance of oscillations remains to be investigated.

Traditional methods such as spectral principal component analysis (PCA), power spectral color map, and spectral nonnegative matrix factorization do not take into account the time-varying nature of these oscillations. The time-varying nature of oscillations can play a vital role in root-cause diagnosis. In an isolated work, Matsuo et al. (2004) explore the T–F spectrum obtained by a wavelet transform to diagnose the root-cause of oscillations in pressure and temperature loops followed by a reevaluation of the performance after remedial actions.

In the work by Selvanathan and Tangirala (2009), the authors propose the use of CWT to distinguish between the different sources of poor loop performance, specifically zooming into the model-plant mismatch (MPM) as the possible cause. MPM can arise due to mismatches in gain, delay, and timeconstants. The cross-wavelet transform (XWT) is the extension of the classical cross-spectrum to the time-scale plane (Grinsted et al., 2004; Torrence and Compo, 1998).

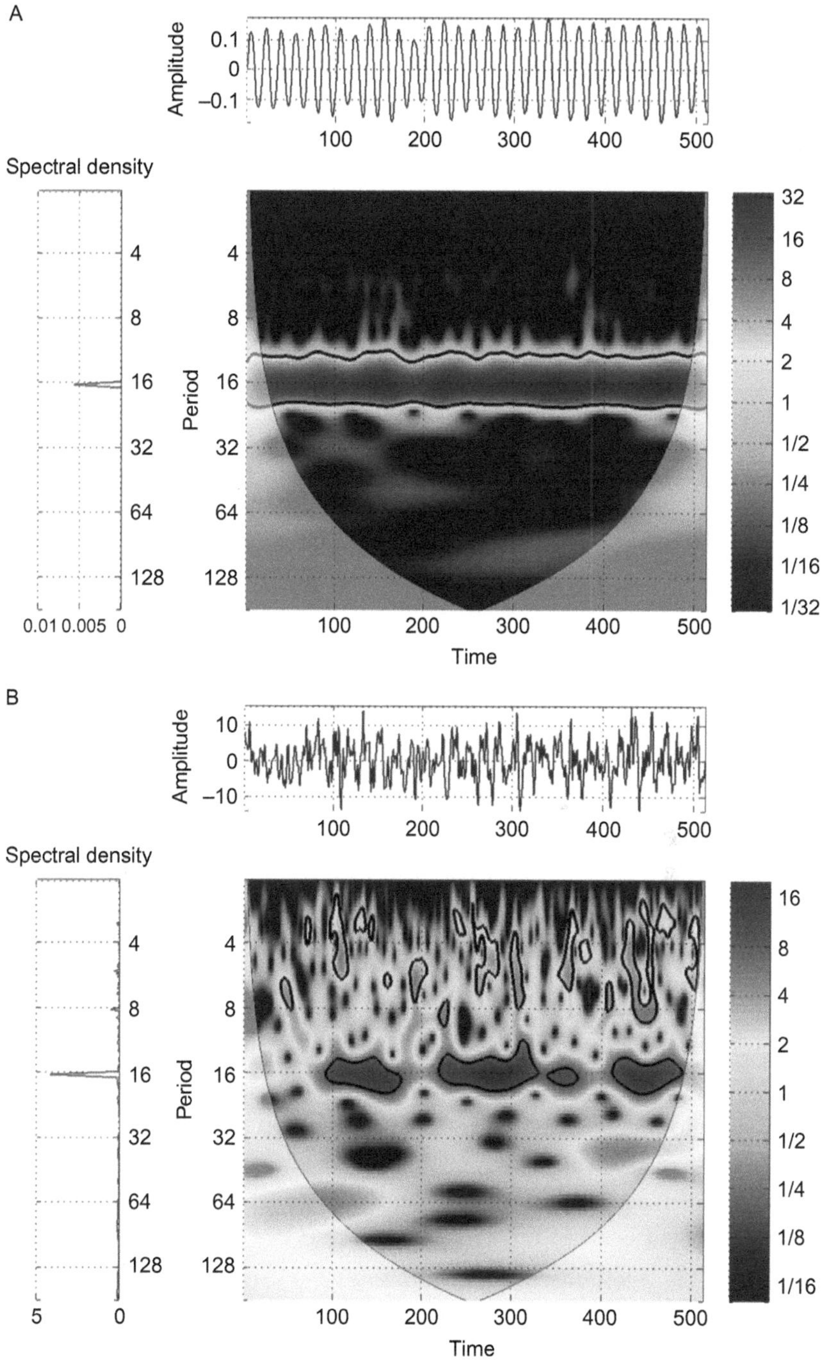

Figure 3.18 Scalogram of measurements reveal the time-varying nature of oscillations in control loops of an industrial process. (A) CWT of a downstream measurement and (B) CWT of an upstream measurement. (For color version of this figure, the reader is referred to the online version of this chapter.)

$$W_{yu}(\tau,s) = W_y(\tau,s)W_u^*(\tau,s). \quad [3.56]$$

Following the analogy, the cross-wavelet spectrum (XWS) is simply the $|W_{yu}(\tau,s)|^2$. A normalized version of XWS is the wavelet coherence (normally abbreviated as WTC) (Grinsted et al., 2004; Maraun and Kurths, 2004), defined as

$$\text{WTC}_{yu}(\tau,s) = \frac{\frac{1}{s}|W_{yu}(\tau,s)|^2}{\sqrt{\frac{1}{s}|W_y(\tau,s)|^2\frac{1}{s}|W_u(\tau,s)|^2}} \quad (3.57)$$

Practical computations of WTC involve the use of a smoothed transform

$$S(W(t,a)) = S_{\text{scale}}(S_{\text{time}}(W(\tau,s))) \quad (3.58)$$

where the smoothing depends on the wavelet. With Morlet wavelets,

$$S_{\text{time}}|_s = \left(W(\tau,s)*c_1 e^{-t^2/2s^2}\right)\Big|_s \quad S_{\text{scale}}|_t = (W(\tau,s)*c_2\text{II}(0.6s))|_t \quad (3.59)$$

where c_1 and c_2 are suitable normalization constants and II is the rectangle function.

The XWT, WPS, and WTC capture the temporal changes in the cross-spectrum, cross-spectral density, and coherence between the input and output of an LTI (and a linear time-varying (LTV)) system. Selvanathan and Tangirala (2009) and Sivalingam and Hovd (2011) exploit the behavior of the magnitude ratios and phase difference between the XWTs of the input with the model output and process response, $W_{\hat{y}u}(\tau,s)$ and $W_{yu}(\tau,s)$, respectively, to diagnose the source of MPM in model-based control schemes. In addition, it is also able to identify the actuator nonlinearities and oscillatory disturbances as a possible source of performance degradation. It is also argued that the XWT provides an edge over the traditional Fourier spectrum of analyzing valve limiting cycles (valve stiction) in that they manifest as discontinuities in addition to the usual harmonic signatures.

As an example, to distinguish between the gain mismatch and an oscillatory disturbance as the source of oscillations, the phase difference for the former source is zero, while for the latter, it is nonzero. Moreover, gain mismatches also cause the ratio of magnitudes of XWT to deviate from unity. Figure 3.19A shows the XWTs $W_{\hat{y}u}$ and W_{yu}, respectively. Below, Fig. 3.19B shows the magnitude ratio and phase difference at the higher frequency, which is known to be due to a gain mismatch. The diagnostics correctly indicate the source of oscillation. In addition, these plots reveal that

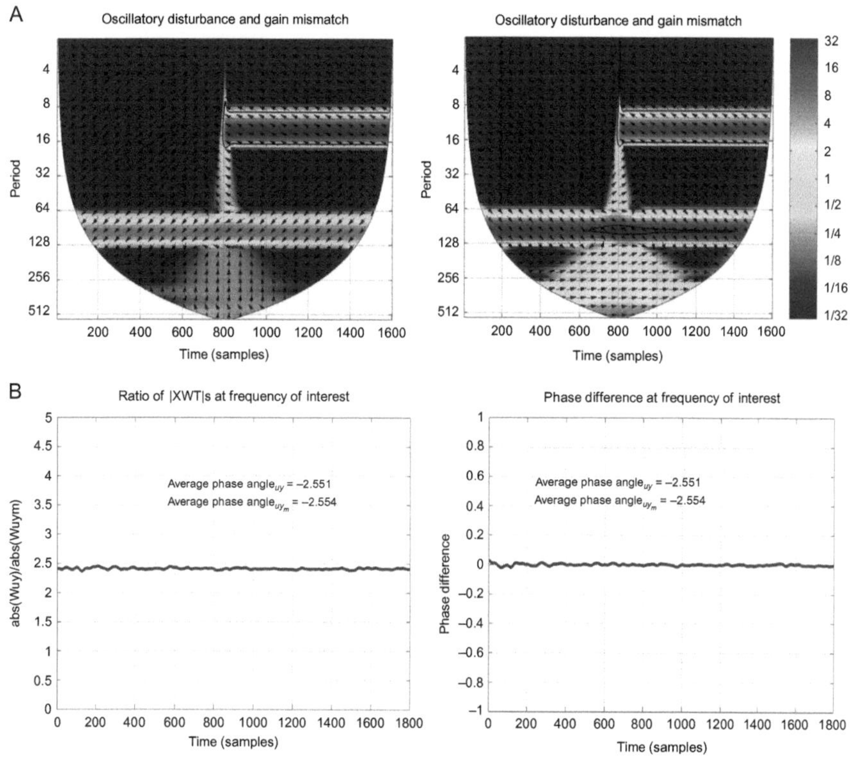

Figure 3.19 Magnitude ratio and phase difference of XWTs are able to distinguish between the sources of oscillation in a model-based control loop. (A) W_{yu} and $W_{\hat{y}u}$ (color: intensity, arrows: phase) and (B) $|W_{yu}(\tau, s)|/|W_{\hat{y}u}|$ and $\angle W_{yu} - \angle W_{\hat{y}u}$ at frequency of interest. (For color version of this figure, the reader is referred to the online version of this chapter.)

the oscillations due to gain mismatch commenced only midway, whereas the oscillatory disturbances persisted throughout the period of observation.

The authors do not provide any statistical tests for the developed diagnostics. Further, a quantification of the valve stiction from the signatures in XWT is missing and potentially a topic for study.

The input–output delay (matrix for MIMO systems) is a critical piece of information in identification and CLPM. Several researchers have attempted to put to use the properties of WT and XWT for this purpose. In a simple approach by Ching et al. (1999), cross-correlation between denoised signals using dyadic wavelet transforms and a newly introduced thresholding algorithm is employed. The method is shown to be superior to traditional cross-correlation method but can be sensitive to threshold. The CWT and wavelet analysis of correlation data have been proved to be more effective for delay estimation as evident from the various methods that have evolved in the past two decades (Ching et al., 1999; Ni et al., 2010; Tabaru, 2007). This should be expected due to the dense sampling of the scale and translation parameter in CWT in contrast to DWT.

Preliminary results in delay estimation using CWT were reported by Tabaru and Shin (1997) using a method based on locating the discontinuity point in the CWT of the step response. The method is sensitive to the presence of noise. Further works exploited other features of CWT. Tabaru (2007) presents a good account of related delay estimation methods, all based on CWT. The main contribution is a theoretical framework to highlight the merits and demerits of the methods. Inspired partly by these works, Ni et al. (2010) develop methods for estimation of delays in multi-input multioutput (MIMO) systems—a challenging problem due to the confounding of correlation between the multiple inputs and the output in the time-domain. The work first constructs correlation functions between CWTs of inputs and outputs of a MIMO system. The key step is to locate nonoverlapping regions of strong correlations between every input–output pair in the T–F plane. Underlying the method is the premise that, where the multivariate input–output correlations are confounded up in time-domain, there exist regions in the T–F plane in which the correlations (between a single output and multiple inputs) are entangled. Consequently, an $m \times m$ MIMO delay estimation problem can be broken up into m^2 SISO delay estimation problems. Although bearing resemblance to the work by Tabaru (2007), the method is shown to be superior and more rigorous. Applications of the method to simulated and pilot-scale data demonstrate its effectiveness. Promising as much the method is it rests on a manual determination of uncorrelated regions. The development rests on the assumption of open-loop conditions. Extensions to closed-loop conditions may be quite involved, particularly the search of regions devoid of confounding between inputs and outputs.

In general, XWTs have been used to analyze phase-locked oscillations (Grinsted et al., 2004; Jevrejeva et al., 2003; Lee, 2002) in climatic and geophysical time series. Both XWT and WTC are bivariate measures. However, a work by Maraun and Kurths (2004) showed that WTC is a more suitable measure to analyze cross-correlations rather than XWT. This is not a surprising result since it is well known that classical coherence is a better suited measure rather than classical cross-power spectrum because the former is a normalized measure (Priestley, 1981). In a recent interesting work, Fernandez-Macho (2012) extend the concepts of XWT to multivariate case deriving new measures known as wavelet multiple correlation and wavelet multiple cross-correlation. These statistics measure correlations in a multivariable process at different scales. The tools potentially have applications in CLPM and identification of multiscale systems.

6.1.2 Modeling

The multiresolution property in the time-scale space of wavelets has been the primary vehicle for modeling multiscale systems. A rigorous formalization of the associated ideas appears in the foundational work by Benveniste et al. (1994) where models on a dyadic tree are introduced. The main outcome is a mechanism or a model that relates signal representations at different scales. A set of recursive relations that describe evolution of system from one scale to the other are developed. Essentially, the model works with coarse to fine prediction or interpolation with higher resolution details added by a filter, coloring a white noise process while going from one scale to next fine scale. The structure admits a class of dynamic models defined locally on the set of nodes (given by scale/transition pairs) and evolving from coarse to fine scales. In doing so, the authors propose the filtering-and-decimation operation for multiscale systems as the equivalence of z-transform used for single-scale LTI systems. Concepts of shifts and stationarity for multiscale systems are redefined. Ideas from this work were later generalized to the data fusion and regularization in Chou et al. (1994). A particular adaptation of the multiscale theory to model-predictive control (MPC) of multiscale systems was presented by Stephanopoulos et al. (2000). Models on binary trees arising from a dyadic WPT are used. It is shown that the computations of the resulting MPC optimization problems can be parallelized across scales. Multiscale MPC application to a batch reactor appears in a work by Krishnan and Hoo (1999). Practical applications of this form of multiscale theory though are very limited primarily due to the mathematical and computational rigor. A major requirement of these methods is the availability of a first principles description of the process.

Over the past decade, a number of ideas have sprung up for identification using wavelets. Kosanovich et al. (1995) introduce the Poisson wavelet transforms (PWT) for identification of LTI systems from step response data. The PWT is a transform of the 1-D signal to the 3-D space characterized by two continuous parameters, τ and β, and one discrete parameter, n (reference). For any signal $x(t)$, the PWT is defined as

$$(W_n x)(\tau,\beta) = \frac{1}{\sqrt{\beta}} \int_{-\infty}^{\infty} f(t)\psi_n\left(\frac{t-\tau}{\beta}\right) \mathrm{d}t \qquad [3.60]$$

$$\psi_n(t) = p_n(t) - p_{n-1}(t) = \begin{cases} \dfrac{(t-n)t^{n-1}\mathrm{e}^{-t}}{n!} & t \geq 0 \\ 0 & t < 0 \end{cases} \quad p_n(t) = \frac{t^n \mathrm{e}^{-t}}{n!} \quad [3.61]$$

where $n \in \mathbb{Z}^+$, and $p_n(t)$ is the Poisson distribution function.

When applied to identification from step response data, PWT essentially decouples the effects of delay and time-constant, which are otherwise entangled in time-domain. This idea is well known in the frequency-domain identification (of LTI systems) literature. PWT offers an improved separability in the effects of dynamics and delays and significantly enhances the estimation of the respective parameters from noisy data by exploiting certain relationships across different values of the discrete parameter n. Ramarathnam and Tangirala (2009) offer correct expressions for these relations and present a systematic procedure for parameter estimation. A major drawback of the PWT-based method is that the applicability is practically limited to first- and second-order systems notwithstanding the theoretical possibility of accommodating higher-order systems.

Wavelets are best utilized when applied to identification of linear/nonlinear time-varying (LTV) systems, which exhibit multiscale behavior. A major challenge in identification of LTV systems is the large number of parameters that have to be estimated. The CWT-based TFR of output and input can be used for reducing dimensionality of the parameter vector Shan and Burl (2011). The methodology rests on the definition of a time-frequency response or a time-frequency representation (TFR)

$$\mathrm{TFR}(j,m) = \frac{W_{(j,m)}[y(t)]}{W(j,m)[u(t)]} \quad [3.62]$$

along the lines of the classical frequency response for LTI systems (Ljung, 1999; Proakis and Manolakis, 2005). Using this measure, scales that are most "informative" (sensitive to the unknown parameters), "noise-free" (good SNR), and "efficient" (minimal sample correlation) are determined. The scale selection is the backbone for dimensionality reduction. Three different criteria to select scales with these features are proposed and evaluated. In addition, an adaptive algorithm to turn on scale selection at deserving time instants to minimize computational workload is also proposed. Although not explicitly stated, this is equivalent to the assumption of local invariance in the T–F plane. A nonlinear least squares estimator that minimizes the sum-squared prediction-errors is used for parameter estimation. The method is demonstrated to be effective for abrupt and very slow changes in parameters. Theoretically, the method is effective in tracking time-variations of the parameters but has two possible shortcomings: (i) the drastic increase in computational burden for systems characterized by frequent changes in parameters and large number of parameters and (ii) it is not straightforward to design metrics that can select scales with the aforementioned characteristics.

CWT finds applications in modeling of mechanical and aerospace engineering systems. A recent work (Xu et al., 2012) narrates a brief summary of related methods while motivating the development of a wavelet-based state-space method for tracking parameter changes in mechanical LTV systems. A wavelet packet-based method is put forth by Paivaa et al. (2006) to model LTI systems. The procedure is once again to decompose the input and output signals into frequency subbands and apply an Akaike information criterion (AIC) (Akaike, 1968; Ljung, 1999) like measure, the generalized cross validation index to achieve parsimony of scales while not compromising on accuracy.

The collective works of Nounou (2006) and Nounou and Nounou (2005, 2007) present ideas for developing multiscale empirical models, namely, the multiscale autoregressive exogenous, multiscale finite impulse response, and multiscale Takagi-Sugeno fuzzy models. A single idea spans these works, that of decomposing data onto different scales, developing separate models on relevant scales and selecting a single model using a prespecified criterion. These methods demonstrate their superiority over single-scale methods but do not fully exploit the advantages of a multiscale decomposition. Selecting the most relevant single scale to represent a process limits the applicability of a multiscale approach.

A comprehensive work by Reis (2009) congregates the existing literature on multiscale modeling of systems on dyadic wavelet trees and single-scale identification techniques to present a sequential procedure for multiscale identification. A salient aspect of this method is that it builds a separate model at every scale that is deemed relevant to the process. Presenting practically implementable ideas, the work demonstrates the significance of each step on three different laboratory and industrial case studies. The role of user-specified parameters, namely, the decomposition depth and the index of selected scales, is studied in detail. A graphical evaluation of the energy decomposition across scales is used to select appropriate values of these parameters. The method deviates from the philosophy of multiscale systems theory suggested by Benveniste et al. (1994) as it does not model the dynamic relationships across scales. Potentially the methodology can be used for modeling several industrial processes while also offering considerable scope for further enhancement and automation.

All the previously discussed methods implicitly or explicitly make an important assumption—the process is stationary in each wavelet subband or scale (local invariance) even while it is nonstationarity or time-varying in the time-domain.

6.2. Wavelets as basis functions for multiscale modeling

The use of wavelets as basis functions brings with it several advantages to the modeling arena. Not surprisingly, numerous works have effectively exploited these advantages, particularly, the sparse representations of signals in wavelet domain (leading to parsimonious models) and to model nonstationary and nonlinear systems (see Juditsky et al., 1995; Sjöberg et al., 1995).

The first class of methods constitutes those approaches in which the classical identification problem is reformulated using orthogonal/biorthogonal wavelets as basis functions, while the model parameters are estimated by minimizing errors in the LS sense (Chang and Qu, 2004; Mukhopadhyay and Tiwari, 2010). An important intermediate step is that of reducing the size of model or the dimensionality of parameters to be estimated (parsimony) by applying appropriate criteria. The assumption of local invariance in the T–F domain is tacitly made.

A discrete-time LTV model in terms of its IR coefficients is given by

$$y[k] = \sum_{n \in \mathbb{Z}} h[k,n]u[k-n] \qquad [3.63]$$

where $h[.,.]$ is the time-varying IR function of the LTV system.

Tsatsanis and Giannakis (1993) model the LTV system by expanding the time-varying IR coefficients $h[k,n]$ in the wavelet basis space using the MRA expansion of Eq. (3.53). It is based on the premise that the time-varying coefficients lend themselves to time-invariant coefficients in the wavelet domain. Parsimony is achieved by selecting a subset of basis using an F-test combined with the AIC. The approach can also be generalized to the use of other basis where the coefficients attain a sparse representation. Wei and Billings (2002) use similar ideas in the modeling of nonlinear and LTV systems using B-spline wavelets as the basis. The prime difference is the use of a different criterion for structure selection, known as the orthogonal forward regression. In a work by Nikolaou and Vuthandam (1998), the multiscale localization properties of wavelet coefficients are explored to build a reduced-order, but for the restrictive class of finite impulse response (FIR) models. The compressed FIR model development relies on a prior qualitative knowledge of the actual length of the FIR model. This method can be treated as a special case of a more general approach discussed below.

Doroslovacki and Fan (1996) take up the more general problem of identification and adaptive filtering of LTV systems using wavelet basis and least mean square (LMS) adaptive filtering algorithm. Moreover, the TVIR is expressed as a linear combination of wavelet basis with time-varying coefficients.

$$h[k,n]=\sum_{I\in\mathbb{Z}}p_I[k]\eta_I[n]\quad h[k,n]=\sum_{I\in\mathbb{Z}}\xi_I[k]p_I[n] \tag{3.64}$$

where $\{\eta_I[.]\}_I$ and $\{\xi[.]\}_I$ are wavelets (or general basis functions) used to expand time-varying response function from input side and output side, respectively. The TVIR of the system is then modeled either from input side or from output side as given below.

$$\begin{aligned}(\text{Input side})\ y[k]&=\sum_{I}p_I[k](\eta_I*u)[k]|y[k]\\&=\sum_{I}\xi_I[k](p_I*u)[k]\ (\text{output side})\end{aligned} \tag{3.65}$$

where $p_I[.]$ are time-varying parameters of the system and $I=(i,j)$ such that i is the shifting and j is the scaling parameter. From either models, it is possible to derive a model structure with *constant* parameters p_{IJ} in the following form,

$$y[k]=\sum_{I}\xi_I[k]\sum_{J}p_{IJ}\left(\eta_J*u\right)[k] \tag{3.66}$$

Modeling of LPTV systems is treated as a specific case in the time-varying framework considering output functions $\{\xi_I(t)\}_I$ to be periodic. In Dorfan et al. (2004), it is shown that a wavelet model is particularly suitable or adaptive identification of linear periodically time-varying systems. While it is claimed in the two aforementioned works that the convergence of the LMS algorithms with wavelets is faster than the LMS FIR algorithm applied in measurement space, it is well known that adaptive LMS algorithms would work only for relatively slowly time-varying processes.

A formal framework of the foregoing ideas is presented by Zhao and Bentsman (2001a,b). Taking a similar stance to that of Doroslovacki and Fan (1996), the LTV model is written as

$$y[k]=\sum_{n=0}^{N_L-1}\sum_{m\in\mathbb{Z}}\sum_{l\in\mathbb{Z}}c_{m,l}\xi_m[k]\eta_l[n]u[k-n], \tag{3.67}$$

which is essentially the expanded version of Eq. (3.66) (restricted to the FIR class). A few differences, but important ones, exist. First, the framework establishes stability conditions for LTV systems and demonstrates convergence of approximation (as more basis functions are included), thereby giving the model a strong mathematical foundation. Second, the adaptive LMS algorithm is not implemented; rather a least squares problem is solved at every instant in time. Third, the modeling approach makes an important assumption—that the LTV system is time invariant over the length of support of the basis functions. The model structure admits a general basis, but the authors recommend the use of spline biorthogonal wavelets.

The modeling ideas in the foregoing works are by far the most generic ones for describing LTV systems. However, some practical concerns remain. The block period over which time invariance is a user-defined parameter, which is most likely decided by trial and error unless some qualitative prior knowledge is available. Solving an LS problem at every instant can be computationally demanding. Although the values of model parameters are updated at every time instant, the approach fails to effectively capture abrupt change in the system such as regime switching in a process. Moreover, linear approximation as suggested by the work may give rise to ill conditioning of estimated IR in certain situations. Finally, the FIR model form.

Extensions of the foregoing methods to multivariable cases are scarce. A related work by Satoa et al. (2007) proposes development of vector autoregressive (VAR) models for multivariable LTV systems. A VAR representation is an extension of the AR model to the multivariable case and is a standard choice for modeling multivariable time series (Lutkepohl, 2005). The work of Satoa et al. (2007) develops the LTV–VAR model using the standard trick, which is to develop a model in terms of wavelet expansion coefficients rather than in signals.

The second class of methods views wavelets as not merely basis functions but also as universal approximators. A method that assumed prominence is the *wavelet network* (see Thuillard, 2000 for a good overview), which naturally accommodates multivariable processes. Seeds of this paradigm were sown in the works by Daugmann (1988), Pati and Krishnaprasad (1992), and Szu et al. (1992), which were contemporaneously formalized in the treatment by Zhang and Benveniste (1992). A neural network is a graphical representation of nonlinear models that use linear combinations of sigmoidal transformations of the input. Similarly, the wavelet network structure uses wavelets as the activation functions, called as *wavelons*. Mathematically, it has the following form:

$$y(\mathbf{x}) = \sum_{i=1}^{N_d} w_i \psi(D_i(\mathbf{x} - \mathbf{t}_i)) + g_0 \qquad [3.68]$$

where D_i is a dilation matrix built from dilation vectors and $\psi(.)$ is the wavelet function. Observe that the network admits a vector signal. Zhang and Benveniste develop the necessary multidimensional wavelet theory. Comparing Eq. (3.68) (barring g_0) with 21, one interprets the wavelet network to be the inverse wavelet transform represented using a neural network architecture with wavelets as activation functions. A distinctive feature of these networks that makes them attractive is the availability of a learning algorithm that adaptively determines the set of dilations and translations necessary for a given dataset. Further, the flexibility of the network in Eq. (3.68) can be enhanced by rotating the data prior to dilation. The rotation assists in modeling along certain directions of interest (such as "axes of maximal information") in the data. The network in Eq. (3.68) then admits a rotation matrix R_i

$$y(\mathbf{x}) = \sum_{i=1}^{N_d} w_i \psi(D_i R_i(\mathbf{x} - \mathbf{t}_i)) + g_0 \qquad [3.69]$$

Two of the possible ways of accommodating rotational information are either by using PCA (Jackson, 1980; Wold et al., 1987) of the original data or by using curvelets (Ma and Plonka, 2010) in place of wavelets.

Zhang and Benveniste (1992) also propose an alternative network known as the wavelet decomposition network (WDN), which is somewhat of a misnomer. The approach consists of performing a wavelet decomposition up to a user-defined depth, coefficients from which feed into a neural network. It is essentially a nonlinear model in the coefficient space. These classes of methods can be viewed as an extension of the aforementioned works to the nonlinear case. Compared to the wavelet network, the WDN carries significantly lesser flexibility and features. Not surprisingly, applications of this network have been very limited.

When the dilations and translations are restricted to the dyadic scales and proportional shifts as in DWT, the wavelet networks acquire the name *wavenets*. This term was coined by Bakshi and Stephanopoulos (1993) in an independent approach. The orthogonal basis functions enable the wavenets to provide a multiresolution approximation of the input–output mapping. The wavenet distinctly differs from the wavelet network of Zhang and Benveniste (1992) in the way it learns. The learning algorithm for the wavenet is noniterative and hierarchical, whereas the wavelet networks learn iteratively through a backpropagation algorithm.

A variant, but subset of wavelet networks, namely, the fuzzy wavelet networks or fuzzy wavenets, was formulated by Thuillard (1999). Wherein the methodology is to engage wavelet scaling functions as membership functions in the Takagi–Sugeno model (Takagi and Sugeno, 1985) for fuzzy rules. Not all scaling functions qualify to be membership functions—they should possess symmetry, be positive everywhere, and have a single maxima. Spline wavelets (scaling functions) are good candidates for this purpose.

Several adaptations of the wavelet networks, wavenets, and their variants have been developed (Aadaleesan et al., 2008; Srivastava et al., 2005; Tzeng, 2010; Wei et al., 2010; Zekri et al., 2008) over the past decade. A noteworthy extension is the combination of wavelet networks with orthonormal basis functions (OBFs) (Aadaleesan et al., 2008). The motivating factor is that wavelet networks are effective in modeling only static nonlinearities while OBFs are capable of representing almost all types of linear, causal, and stable systems. The OBFs a general category of filters that include FIR, Laguerre, and Kautz filters as special cases (Ninness and Gustaffson, 1997). While the concatenation of OBFs with a wavelet network is worthy, additionally placing a Wiener or Hammerstein model in series with the OBF-wavenet is contentious. It is based on the argument that a wavelet network cannot effectively and parsimoniously handle linearities or mild nonlinearities. This argument lacks conviction fundamentally since it contradicts the universal approximation abilities of a wavelet network and is also in contrast to the properties of wavelet coefficients.

Wavelet networks and their extensions have been applied quite successfully in modeling and control applications (cf. Aadaleesan et al., 2008; Chang et al., 1998; Katic and Vukobratovic, 1997; Safavi and Romagnoli, 1997). However, wavelet networks (and wavenets) remain far from being fully explored to their potential. The learning algorithms of wavelet networks can be very sensitive to the initial guesses of the unknowns. The crucial decision on the number of wavelets and the type of wavelets to be used rests with the user. A stepwise procedure is detailed in Sjöberg et al. (1995). The authors coin the term *constructive approach* to the method of selecting wavelet bases and appropriate dilations from data. Of particular concern is the ability to construct a multidimensional wavelet as the dimension becomes large. Several studies report the impact of these decision variables on the complexity and quality of developed networks.

Sureshbabu and Farrell (1999) take a different stance on the use of wavelets as universal approximators in their approach to nonparametric identification of nonlinear systems using wavelets. They argue that a network like structure may not be necessary with a careful choice of the depths and the basis functions. However, a convincing demonstrating is lacking. The applicability is quite limited due to the conservative assumptions made on the nonlinear nature. Further, only the univariate case is considered. Extensions to the multivariable case do not appear to be straightforward.

In an interesting parallel to the wavelet network concepts, Lu et al. (2009) deploy wavelets as kernel functions in a support vector regression (SVR) framework. Using theoretical comparisons, it is argued that the wavelet-kernel SVR using linear programming optimization represents an optimal wavelet network.

Another powerful class of models combines wavelet-based expansions of nonlinearities with polynomial models in the nonlinear autoregressive moving average exogenous (NARMAX) setting (Billings and Wei, 2005), as follows

$$y(t) = f(x(t)) = \underbrace{f^P(\mathbf{x}(t))}_{\text{Polynomial model}} + \underbrace{f^W(\mathbf{x}(t))}_{\text{Wavelet model}} + \underbrace{f^E(\epsilon(t))}_{\text{Error model}} + \mathrm{e}(t) \quad [3.70]$$

where $\mathbf{x}(t)$ is the vector of regressors containing *past* outputs and inputs and ϵ is the vector of past errors, both up to a user-specified lag. The wavelet component of the WANARMAX representation admits a multiresolution approximation of the output. A recommended choice of wavelets (scaling functions) is the *B*-spline wavelets. Equation (3.70) can be cast into a linear-in-parameters form. Model parsimony (selection of relevant terms) is achieved by a hybrid matching pursuit (Mallat and Zhang, 1993) and the orthogonal least squares algorithm. The development presents little discussion or argument on the inclusion of a polynomial term in the presence of a wavelet approximation term. Further, the WANARMAX in principle can effectively model a wide range of processes and possesses similar capabilities as the wavelet networks. However, the computational costs with these classes of models can assume serious proportions. The orders of the outputs and inputs, as in the classical identification case, have to be chosen by trial and error.

6.3. Wavelets as multiscale filters for modeling

Using wavelets as multiscale filters for modeling generally rests on the idea of decomposing the outputs and/or inputs into frequency subbands using suitable wavelet filters and then applying predetermined or adaptive criteria to include only the relevant subbands in reconstruction. The reconstructed signals are then used for developing models.

Palavajjhala et al. (1995) used the criteria of maximizing SNR to select the relevant scales for identification. Adopting the multirate band-pass filtering property of wavelets, Carrier and Stephanopoulos (1998) floated ideas for control relevant identification by building reduced-order models for LTV and nonlinear systems. The inputs and outputs are first filtered into frequency subbands to subsequently identify the frequency bands (scales) that are crucial to a stable and efficient control of the process. The knowledge of crossover frequency is used to determine the relevant scales. However, the concept of a single crossover frequency for a multiscale system is ambiguous.

Chang and Qu (2004) formulate a $\|.\|_1$-norm penalized least squares problem to achieve sparseness in wavelet domain for the identification of partially linear models. This is intuitively equivalent to jointly denoising the coefficients and identifying the model using cleaned signals.

6.3.1 Unified perspective

It is worthwhile noting that the apparently different approaches in Sections 6.1–6.3 can all be brought under a single umbrella of methods that extend two standard ideas of classical identification, namely, (i) prefiltering (in the Fourier domain) and (ii) identification in a transform basis, to prefiltering in the T–F plane using a wavelet basis. The wavelet basis naturally accommodates a bank of filters. Importantly, the wavelet filters set the platform for a hierarchical framework in the time-scale space. A particular method may incorporate one or both of these extensions. Just as prefiltering can be viewed as regularization in the classical identification framework, the selection of relevant bands (scales) can be viewed as an equivalent of regularization in the projection coefficient space.

7. CONSISTENT PREDICTION MODELING USING WAVELETS

7.1. Introduction

The review of algorithms in the foregoing Section 4.7, particularly that of Section 6.2, suggests refinements or alternatives for modeling of linear/nonlinear TV systems in two respects. The first one is that the existing methods (of Section 4.7) solve the identification problem in the time-domain even as the signal representation is in a new basis. Thus, they do not fully exploit the separability achieved in the coefficient space. Second, determining the model terms to be retained can be done in a more efficient manner using the ideas of consistent estimation. Keeping these two points, an alternative modeling approach based on ideas in Mukhopadhyay and Tiwari (2010) is explored. Preliminary results of this approach were presented in Mukhopadhyay et al. (2010).

The alternative approach is based on the notion of *consistent prediction* (an extension of the concept of consistent estimation) and undecimated dyadic transform (or MODWT).

The proposed approach requires no assumption of local time invariance. Three distinct features of the alternative approach can be observed: (i) *developed model is built on projection coefficients* (thereby exploiting the nice separability and decorrelated properties of the coefficients), (ii) consistent prediction of output signal coefficients (thereby eliminating noise effectively), and (iii) subband identification (one that captures the differences in the frequency responses over different bands). Last, the wavelet basis is spline biorthogonal wavelet basis, carrying with it several advantages. An advantage deserving attention is that using splines as basis, *direct weighted addition of projections* in approximation space can be used for consistent output predictions. Further, it can be shown that the solution seeking local fit in approximation space does not necessarily require the assumption of strict orthogonality.

The consistent prediction is defined in a similar way as consistent estimation as follows.

Definition

A consistent prediction is that prediction whose wavelet representation is identical to that of the signal component of the measurement in wavelet domain.

The method of parameter estimation proposed in this work produces nonlinear approximation (Mallat, 1999) and primarily checks the local consistency of the estimate with output signal for a determinable minimum memory solution in wavelet domain.

Although derived through a different route, this parametric identification approach bears similarity to the method of Shan and Burl (2011). A notable benefit of the proposed method is that it identifies a system truly in multiresolution spaces, thus also computationally being superior. An elegant algorithmic implementation is also provided for the proposed method.

A positive by-product of this work is the development of a method that can efficiently isolate multiple timescales in a linear time-invariant system, a problem that has been of interest and challenge in identification and control (Tiwari et al., 2000). A fine separation of timescales is not possible using wavelet transforms. The reason for this can be understood as either due to the duration–bandwidth principle or equivalently that wavelets cannot diagonalize differentiation operators.

The efficacy of the technique is demonstrated in two case studies by modeling complexities (integrating effect, multiple time-scale behavior, and nonlinearity) using time-varying wavelet models.

7.1.1 A time-varying model with spline biorthogonal wavelets as basis

Following the line of thought in Doroslovacki and Fan (1996), Tsatsanis and Giannakis (1993), and Zhao and Bentsman (2001b), the *one-step ahead predictor* for the LTV system can be written as

$$\hat{y}(t) = \sum_i p_i(t)(\kappa_i * y)(t) + \sum_i q_i(t)(\gamma_i * u)(t) \qquad [3.71]$$

where κ_i, $i=1, 2, \ldots M$ and γ_i, $i=1, 2, \ldots N$ two different finite length basis functions for projecting finite length outputs and inputs, respectively, in the approximation space. The subscripts denote discrete indices of the basis functions.

The formulation in Eq. (3.71) differs from the ones used in Doroslovacki and Fan (1996) and Zhao and Bentsman (2001a,b) in that a TVARX model is adopted in place of the restrictive FIR structure. Expanding the coefficients on the output *synthesis* (dual) basis functions

$$p_i(t) = \sum_l p_{il}\tilde{\kappa}_l(t) \quad q_i(t) = \sum_l q_{il}\tilde{\kappa}_l(t)$$

the discrete-time one-step-ahead prediction model can then be expressed in terms of real-valued constant parameters p_{il}s and q_{il}s. For spline biorthogonal wavelet basis with symmetry,

$$\hat{y}[k+1] = \sum_k \sum_l p_{kl}\langle \kappa_k, y\rangle \tilde{\kappa}_l[k] + \sum_k \sum_l q_{kl}\langle \gamma_k, u\rangle \tilde{\gamma}_l[k] \qquad [3.72]$$

Observe that when κ_i and γ_i are *sinc* functions, the convolutions in Eq. (3.72) select time samples of input and output and the model reduces to the classical ARX type model.

7.2. Consistent output prediction-based methodology

Denote the shifted version of measurement $y(t+T)$, as $y_s(t)$, where T is the sampling time. The measurement $y[k+1]$ can be then expressed in terms of projections of shifted version of the measurement $y_s(t)$ onto κ_l

$$y[k+1] = (y_s)[k] = \sum_l \langle \kappa_l, y_s \rangle \widetilde{\kappa}_l[k] \qquad [3.73]$$

where as usual $\widetilde{\kappa}_l$ denotes the reconstruction (dual) wavelet.

Minimum error solution in the least squares sense is obtained by minimizing the error functional,

$$J = \sum_k (y[k+1] - \hat{y}[k+1])^2 = \sum_k \epsilon^2[k], \qquad [3.74]$$

where

$$\epsilon[k] = \sum_l \langle \kappa_l, y_s \rangle \widetilde{\kappa}_l[k] - \sum_l \sum_k p_{kl} \langle \kappa_k, y \rangle \widetilde{\kappa}_l[k] - \sum_l \sum_k q_{kl} \langle \gamma_k, u \rangle \widetilde{\kappa}_l[k]$$

The attempt here is to identify a system by p_{kl}'s and q_{kl}'s completely in wavelet domain as shown below. This is in deviation from traditional approaches of identification where model parameters are estimated by minimizing error function defined in time. If orthogonal wavelets are used, energy of the signal is equal by Parseval's relation in both time and wavelet domain and hence error minimization in LS sense in either domain would give the same solution. In fact, this has been exploited in a few works on identification using nonwavelet orthonormal functions (Patwardhan and Shah, 2006; Patwardhan et al., 2006). *However, when biorthogonal wavelets are used, the error minimizations need not be identical.*

In the existing methods, it is assumed that the parameters p_{kl} and q_{kl} are time invariant over each scale. Relaxing this condition, the system remains time varying in transform domain. Then it is wiser to minimize local error instead of global error. Given the wavelet basis and the chosen thresholds for the modeling exercise, identification solution is optimum (in LS sense).

7.3. Proposed solution

Using the undecimated wavelet transform, the index k represents the original sampling grid. The error in time $e[k]$ can be written in terms of errors in wavelet domain $\epsilon_W[l]$

$$\begin{aligned} \epsilon[k] &= \sum_l \left[\langle \kappa_l, y_s \rangle - \sum_k [p_{kl} \langle \kappa_k, y \rangle + q_{kl} \langle \gamma_k * u \rangle] \right] \widetilde{\kappa}_l[k] \\ &= \sum_l \epsilon_W[l] \widetilde{\kappa}_l[k] \end{aligned} \qquad [3.75]$$

The parameters are estimated as usual by setting the partial derivatives to zero. For instance,

$$\frac{\partial J}{\partial p_{kl}} = 2\sum_k e[k] \langle \kappa_k, y \rangle \widetilde{\kappa}_l[k] = 0 \qquad [3.76]$$

It can be seen from Eq. (3.76) that a solution is obtained by either setting error in time $e[k]=0$ or projection coefficient $\langle \kappa_k, y \rangle = 0$.

Remark

Since $\widetilde{\kappa}_l$ spans the output error space, from Eq. (3.75), it follows that $e[k]=0 \Rightarrow \epsilon_W[l]=0 \quad \forall l=k$. Forcing $\epsilon_W[l]$ to zero at all values of $l=k$ implies forcing the predictions and the measurements to exactly match *in the wavelet domain*. This is obviously an underdetermined problem. Hence the error is set to zero (in the wavelet domain) only at "significant" values of l, which is determined by a thresholding procedure. The process of estimating parameters such that the predictions match the significant values in the wavelet domain at significant points is the philosophy of consistent prediction. This can also be thought of the classical regularization (penalized minimization, where the objective is to reduce the number of parameters to be estimated by adding a penalty term to the objective function. Essentially, parsimony or sparsity is achieved by virtue of consistent prediction.

Let λ_u and λ_y be two strictly positive values. In penalized minimization, only those wavelet projections of input and output are used which have modulus values more than λ_u and λ_y, respectively. These projections are the significant wavelet projections. Reckoning that $\epsilon_W[l]$ for all $l=k$ is scalar summation of wavelet coefficients only at kth instant, l in the subscript of p and q can be dropped. The solution of consistent output prediction can be written as (Mukhopadhyay and Tiwari, 2010).

$$\begin{aligned} &\langle \kappa_k, y_s \rangle - p_k \langle \kappa_k, y \rangle - q_k \langle \gamma_k, u \rangle = 0, \ \forall k \in \{ I_u : |\langle \gamma_k, u \rangle| \geq \lambda_u \cap I_y : |\langle \kappa_k, y \rangle| \geq \lambda_y \} \\ &p_k = q_k = 0, \ \forall k \notin I_u \text{ and } \forall k \notin I_y \end{aligned} \qquad [3.77]$$

If $\dim(I_u \cap I_y) = M$, the system is identified in M-dim subspace with $(M \ll K)$. At each k, one still needs to find two parameters p_k and q_k from a single equation. An algorithmic solution for the identification of an LTV model is provided in Appendix C. The alternate projection algorithm summarized in Appendix C also reconstructs the consistent output estimate (one-step-ahead prediction) for model testing and cross-validation.

Note that at indices k for which $\langle \kappa_k, y \rangle = 0$, error in time $e[k]$ is not necessarily zero. At all other k, $e[k]$ is exactly equal to zero.

In Appendix B, it is shown that the estimator based on consistent output prediction is unbiased and has error bounds that can be controlled by choice of threshold.

Although this formulation is in general time varying, it is useful to derive an approximate LTI model when it is known *a priori* (from physics) that the process is LTI.

7.4. Demonstration of results and discussion

To demonstrate the methodology, we present a simulation case study with a simple transfer function model of an LTI system.

7.4.1 Case study 1: Transfer function model

Consider a transfer function $T(s)$ of a fourth order system having poles at $s = 0, -1, -8, -9$

$$T(s) = \frac{0.08(s^3 + 6s^2 + 6s + 8)}{s^4 + 18s^3 + 89s^2 + 72s}$$

The dynamic modes of the above system can be grouped into two categories, one containing the fast ones due to poles at $s = -8, -9$ and the other(s) containing slower ones due to poles at $s = 0, -1$. The motivation for modeling with wavelet basis is to isolate these groups as finely as possible.

An LTI model $h(\tau)$ is derived by exciting $T(s)$ with a train of impulses and by consistent prediction of the output as shown in Fig. 3.20A. The identification amounts to estimating parameters of the system function in Eq. (3.77) by assuming the parameters to be constant over each scale j. The parameters p_j given in Table 3.2 show a definite, if not fine, separation of modes in the wavelet model, fast mode indicated by p_0 and the slower ones by p_2, p_3, and p_4. The model is cross validated by using a sinusoidal input and matching actual and predicted output as shown in Fig. 3.20B. It may be noted that decimated wavelet transform naturally isolates slow and fast operating modes with optimum resolution. However, number of parameters indicating a group shall depend on several factors such as width

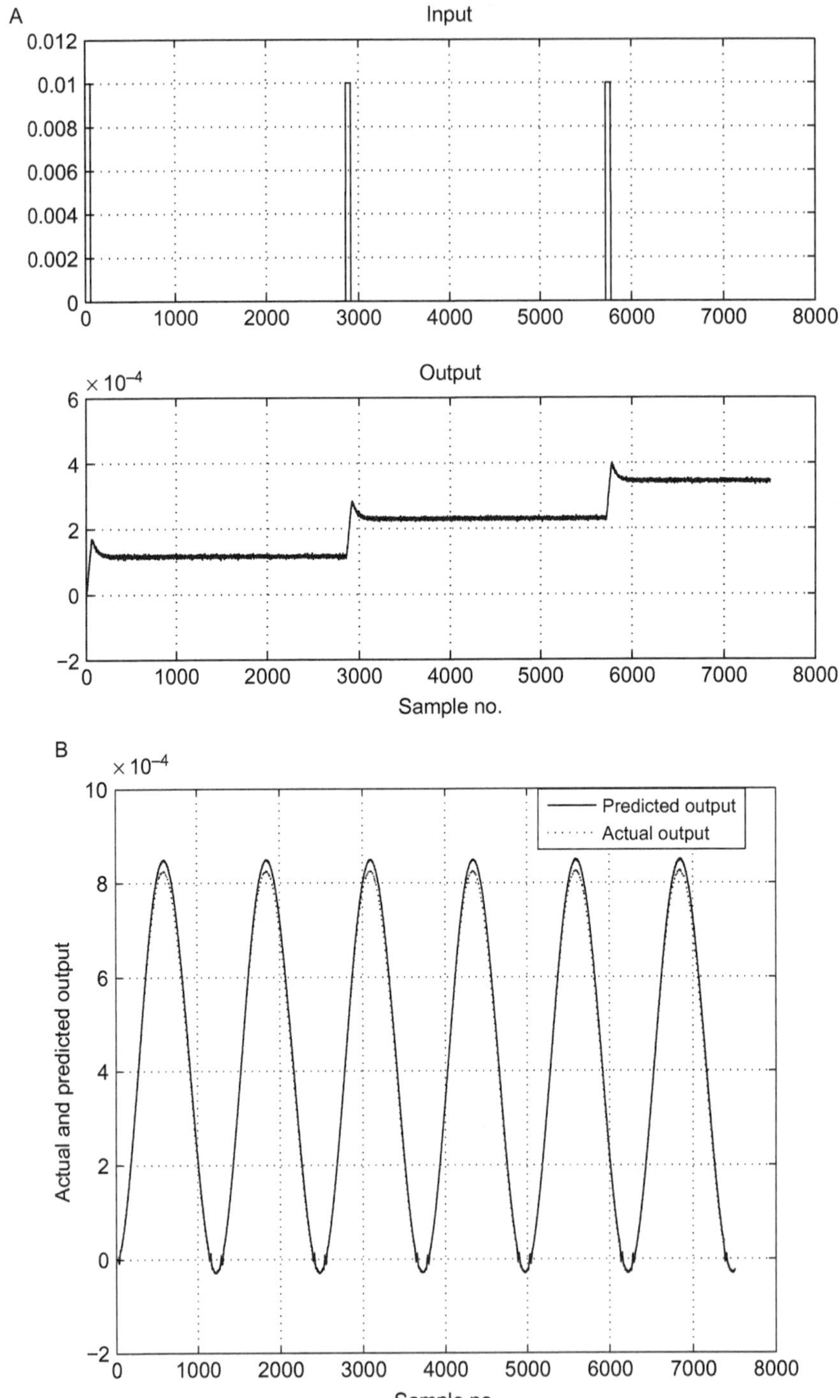

Figure 3.20 Training data and cross-validation for the simulation case study. (A) Training data for identification and (B) cross-validation with sine wave input.

Table 3.2 Estimated parameters of the LTI model

Scale index, j	1	2	3	4	4
$\hat{p}_j$	−0.5	0.4	0.8	1.1	1.0
$\hat{q}_j$ ($\times 10^{-4}$)	−6.1	2.4	−1.1	2.7	−0.3

of the frequency window chosen for analysis vis-a-vis width of the group, sampling frequency, etc.

The proposed technique can be used to efficiently model systems characterized by fast transients superimposed on slowly varying quasi-steady states.

The technique of parameter estimation is further demonstrated by LTV modeling of the liquid zone control system (LZCS) in a large pressurized heavy water reactor (PHWR) (Mukhopadhyay and Tiwari, 2010).

7.4.2 Case study 2: Liquid zone control system

The 540-MW nuclear reactor consists of 14 zone control compartments (ZCC). Control of the reactor power level and the core power distribution is achieved by LZCS through variation of light water levels in the ZCC. Figure 3.21A and B depicts two sets of input–output data collected from a full size LZCS test set-up at 50 ms uniform interval. Input signal is shown as the equivalent desired position of the control valve (CV) in terms of percentage opening (%OPN). The output signal is the level of water expressed as percentage of full scale (%FS). Full-scale level means that the height of the water column is equal to the full height of the ZCC. In the experiments, the water level in each ZCC was regulated by its level controller.

A simple first-order LTI model required for the design of reactor regulating system can be developed from first principles considering the ZCC as a tank. Although the first-order model is adequate for the initial design of control system, simulation needs rigorous models of LZCS needing knowledge of valve design data including the characteristics of its different accessories. In view of such difficulties, developing the model for ZCC water level dynamics employing a suitable method of identification from measurement of input and output is preferred.

The LTV modeling approach due to Zhao and Bentsman (2001a,b), which assumes block time invariance, when applied to the LZCS, failed in cross-validation. This is primarily because the approach precludes nonlinear operation such as thresholding, resulting locally unstable solution. The

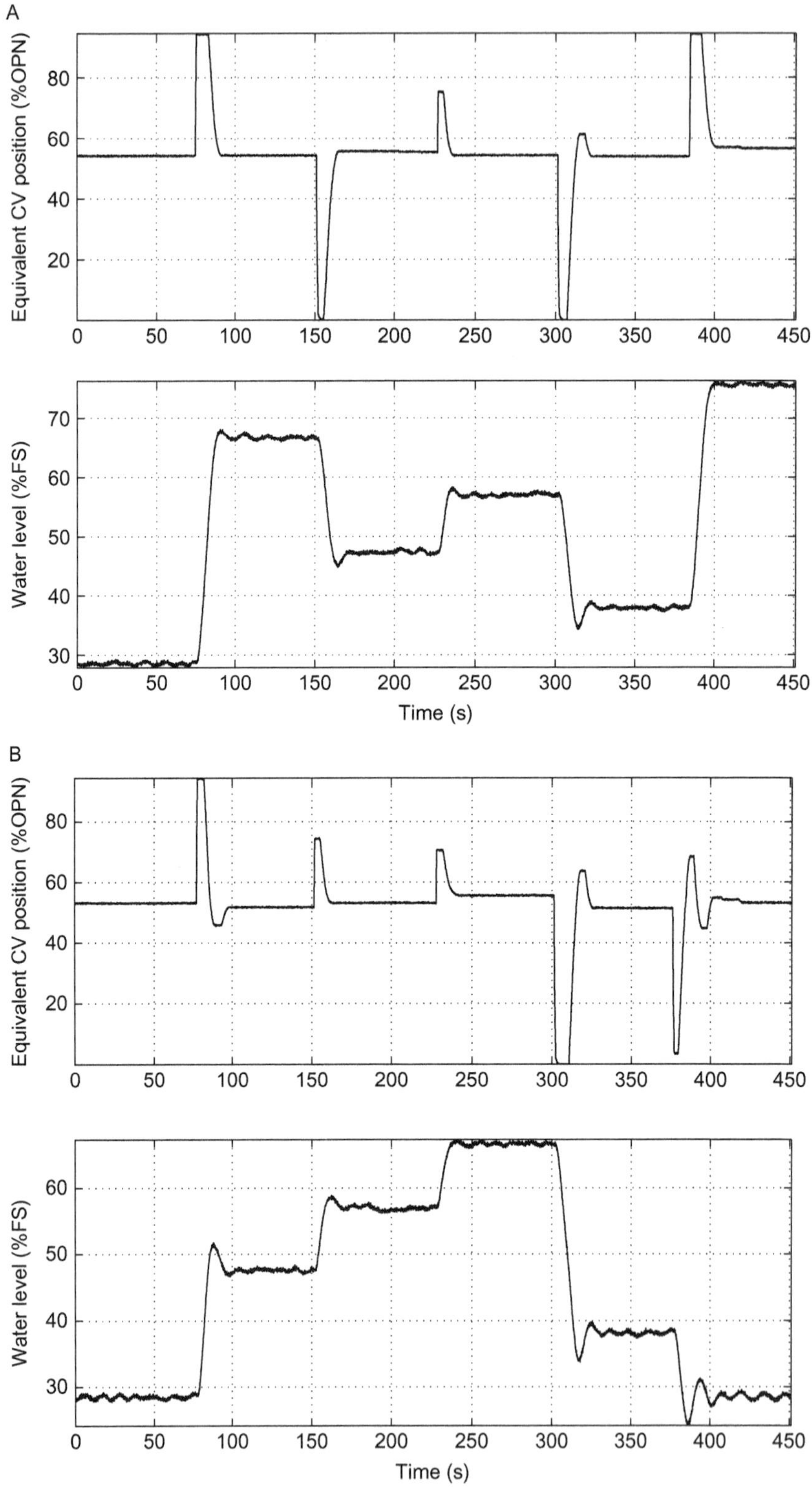

Figure 3.21 Input–output profiles for modeling the LZCS reactor. (A) Input–output profile for training and (B) input–output profile for validation.

instability arises due to ill conditioning of the regressor matrix and invalid assumption of local time invariance failing to model rapid changes in the response.

A LTV model of the LZCS was developed using consistent output prediction with spline biorthogonal wavelets. Two spline biorthogonal wavelets of different orders are used, one for projecting the input and the other for projecting the output. Wavelet RBIO1.5 is used for projecting or analyzing the input. The analyzing scaling function of RBIO1.5 is a box function or box spline of degree zero. Projection of step input on the scaling and wavelet functions of RBIO1.5 shall minimize number of significant wavelet coefficients.

The data of Fig. 3.21A are used for identification of the model, and data from the second experiment (B) is used for validation of the model. The proposed iterative alternate projection algorithm estimates the time-varying parameters at each scale. The reconstructed water level output signal (after error settles to a low value in a few iterations) and actual water level output signal are compared in Fig. 3.22A. A good match is observed between the consistent prediction and the actual output.

The identified model based on the input–output data given in Fig. 3.21A, thus obtained, is now tested also with the input–output data shown in Fig. 3.21B to check if actual output can be predicted. The output in this case is again measured by exciting the CV with a different sequence of steps. The cross-validation result is shown in Fig. 3.22B.

It is known from the physics of the LZCS that the process is only mildly nonlinear and it is worth investigating the performance of a LTI-over-a-scale model (at each scale). The constant value of the parameter at scale j is obtained by averaging the time-varying parameter values at the same scale.

An excellent match is observed in the cross-validation result, between the actual output and the prediction by a subband LTI model (Fig. 3.22B). The match is good in both the transient and steady-state responses, between the model output and the actual output level of the ZCC. It is clear that the use of two different wavelet bases with underlying spline biorthogonal function for modeling input and output reduces the number of wavelet coefficients and gives a smoother approximation in case output is approximated with higher order basis. The results conclusively prove the validity of proposed method of parameter estimation based on consistent output prediction.

7.5. Summary

This section introduced consistent output prediction in a wavelet domain in spline biorthogonal wavelets as an algorithmic solution to least squares

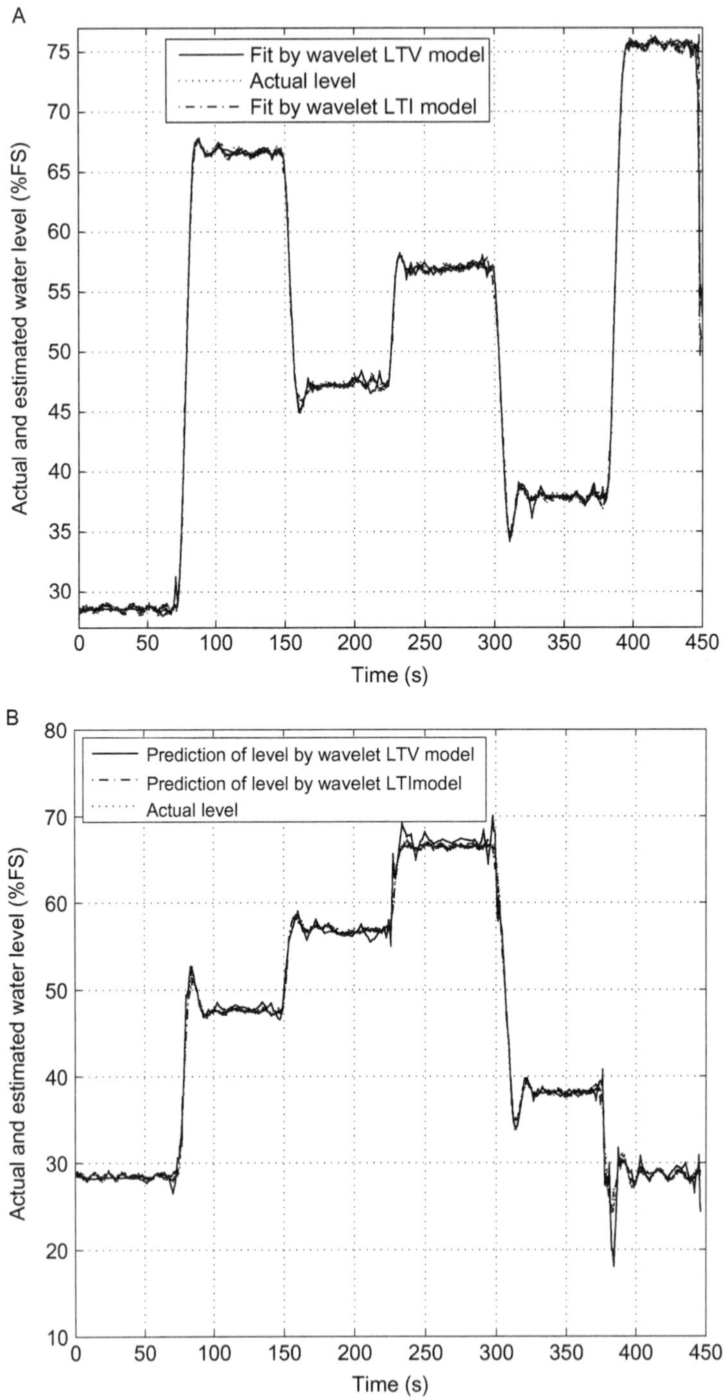

Figure 3.22 Performance of the model on training and test data set. (A) Actual versus predicted levels on training data set and (B) actual versus predicted levels on validation data set.

minimization problem. Penalized minimization of local errors in wavelet domain is used to obtain estimate of system parameters. The algorithm is computationally efficient and as spline biorthogonal wavelets are used, direct weighted summation of projections are permitted and assumption of strict orthogonality is not needed. The theory is validated by means of applications to two case studies (i) modeling of a multiple time-scale system where the timescales were isolated effectively and (ii) identification of the LZCS in a large PHWR where the predictions of the model estimated by the proposed iterative alternate projection algorithm showed excellent agreement with the experimental data in cross-validation.

8. CONCLUDING REMARKS AND FUTURE DIRECTIONS

Wavelet transforms have emerged as efficient and powerful tools for a variety of engineering and scientific applications, particularly for signal estimation, multiresolution approximations, and multiscale analysis. Perhaps the most important reason for their popularity is their ability to hierarchically zoom into (and out of) the information present across scales with mathematical ease. The birth of wavelet transform was a culmination of harmonious efforts by mathematicians, physicists, engineers, and researchers from other fields. Owing to this fact, wavelets have played diverse roles—as basis functions for functional analysis and signal representations, as filter banks for signal processing, as T–F atoms for harmonic analysis of transient signals and feature extraction and as operators in functional analysis and calculus.

The chapter presented an overview of concepts and applications of wavelet transforms to modeling and control of time-varying and, in general, multiscale systems. Chemical engineering curricula across the globe do not comprise courses providing solid foundations in wavelet analysis or signal processing. With this justification and the objective of providing a one-stop resource, a significant portion of this chapter has been devoted to introductory concepts. It is hoped that the tutorial-style introduction of the subject with historical perspectives will help a beginner embrace the subject with ease.

The review presented in this chapter brought together the state-of-art wavelet-based methods available for modeling and CLPM. The focus has been on methods with a broader scope, while excluding those which are highly customized to a particular application. In presenting a critical review of the existing methodologies for modeling and CLPM, the objectives were to (i) provide an awareness of the benefits and limitations of a particular method and (ii) offer ideas for potential extensions and innovations.

A great deal of developments both at the research and pedagogical levels have to occur before these methodologies blossom into practically useful tools for modeling and control of industrial processes. With this vision, some suggestions for future directions are proposed:

1. Mathematically complicated multiscale systems theory and models are likely to draw lesser attention of an engineer as compared to a less complex, but practically implementable solution. For example, what process characteristics demand the use of a wavenet or WANARMAX versus a simpler LTV model? In this respect, works conducting a critical comparison of existing methods, not merely the theoretical but also the practical aspects, are clearly the need of the hour. In such a study, important factors such as computational complexity, scope for automation, and user-friendliness should be given top considerations.
2. An ideally desirable representation of a multiscale system is a model at the finest scale accompanied by a decomposition equation that produces models at coarser scales, that is, the MRA of systems. The classic multiscale systems theory by Benveniste et al. (1994) is centered around this idea. However, in empirical modeling, little or no effort has been placed to ensure that models identified at different scales do belong to the MRA space of a system. The effort until now has been only to identify models separately at each scales. An appropriate integration of these scale-dependent models and an enforcement that a model at a coarser scale falls out of the model at a finer scale is missing. This topic calls for a significant reformulation of the identification problem for multiscale systems.
3. It is known that nonlinear and time-varying systems are characterized by frequency interactions. Models that include cross-band terms (across scales) are suitable for this purpose. Existing methods do not take this into account. Such models may have to include cross-band terms both at the input–output and output–output level.
4. Most methods as observed in this review minimize cost functions in time-domain. As proposed in Section 7, such methods do not fully exploit the properties of wavelet coefficients. Sacrificing orthogonality gives rise to flexibility in modeling. Spline biorthogonal wavelets offer good alternatives as demonstrated in recent works and in the proposed work.
5. The use of wavelets in a state-space framework has been rather limited, virtually negligible in the identification arena. Tremendous opportunities exist in this direction. State-space models could be identified at

multiple scales. A minor but important issue that requires investigation is that of identifiability of these models.

6. In controller design, wavelets have hardly been exploited. An idea that merits exploration is that of control in the wavelet (sub)space, essentially control in a new basis space. A recent paper by Cole et al. (2006) takes this route for active vibration control by reformulating the control problem in terms of the wavelet coefficients. An obvious advantage of controlling in wavelet domain is the enhanced separation between the noise and signal owing to the separability property of wavelet coefficients. Another advantage is that models developed in wavelet domain can be directly employed for control.
7. Engineering problems of today are increasingly multidisciplinary in nature. To be able to borrow ideas from other fields, and apply them to model and control multiscale processes using time-scale analysis tools, researchers necessarily require a strong base in applied signal processing, linear algebra, and applied statistics. A significant step toward realizing this goal would be to include these requirements in the curricula of graduate programs in chemical engineering and allied fields.

In closing, we make two remarks. Wavelet applications to other branches of systems engineering such as monitoring, pattern recognition, and multivariable data analysis are abundant. These adaptations also follow ideas similar to those elucidated for the focus applications in this chapter. On the other hand, there exist niche concepts such as design of custom wavelets using lifting techniques, analysis using multiwavelets, combination of wavelet filters with Kalman filters, etc. Applications based on these concepts appear futuristic with only specialized applications as of today. Second and last, the availability of a plethora of wavelet-based techniques is not indicative of the maturity of the subject by any yardstick, but is rather a strong evidence of the shifts from single-scale to multiscale approaches. These shifts usher a new era in the (data-driven) analysis of multiscale systems. It took nearly one-and-half century for the concept of Fourier transforms to assume an indispensable place. It is envisaged that wavelet transforms shall occupy a similar or even a stronger position on a shorter timescale.

ACKNOWLEDGMENTS

The authors gratefully acknowledge the developers of the software packages, Wavelab, Time–Frequency Toolbox, and WTC Toolbox for their immense generosity in providing their software in an open-source and free environment.

APPENDIX A. PROJECTIONS, APPROXIMATIONS, AND DETAILS

The projection of a vector $\mathbf{x}$ on another vector $\mathbf{v}_i$ is given by

$$\mathcal{P}_{\mathbf{v}_i}\mathbf{x} = \frac{\langle \mathbf{x}, \mathbf{v}_i \rangle}{\|\mathbf{v}_i\|_2^2}\mathbf{v}_i \quad \text{[A.1]}$$

where the notation $\langle , \rangle$ denotes the inner product, specifically here the dot product between two vectors.

A transform involves projection of $\mathbf{x}$ onto a *subspace* V (of the space S to which $\mathbf{x}$ belongs) spanned by a set of basis vectors $\mathbf{v}_i$, $i \in \mathcal{Z}$.

Projections have widespread applications. The discrete-time signal (i.e., due to *sampling*) is a result of the projection of continuous-time signal $x(t)$ onto the *sinc* basis functions according to Shannon's reconstruction formula,

$$x(t) = \sum_k x[k] \operatorname{sin} c\left(\frac{t - kT_s}{T_s}\right)$$

If the basis vectors $\mathbf{v}_i$ are orthogonal, the projection of $\mathbf{x}$ onto a subspace V is the sum of projections onto the individual vectors,

$$\mathcal{P}_V\mathbf{x} = \sum_{i \in \mathcal{Z}} \frac{\langle \mathbf{x}, \mathbf{v}_i \rangle}{\|\mathbf{v}_i\|_2^2}\mathbf{v}_i \quad \text{[A.2]}$$

Thus, orthogonality of basis implies each projection $\mathcal{P}_{\mathbf{v}_i}\mathbf{x}$ carries a new piece of information about the vector $\mathbf{x}$. The new representation of the signal in V is the scalar *coefficient of projection*,

$$c_i = \frac{\langle \mathbf{x}, \mathbf{v}_i \rangle}{\|\mathbf{v}_i\|_2^2} \quad i \in \mathcal{Z} \quad \text{[A.3]}$$

also known as the *transform coefficient* (or simply coefficient). It is a very useful quantity in signal analysis. Orthogonal $\{\mathbf{v}_i\}$s result in each coefficient containing a unique piece of information about $\mathbf{x}$.

If the basis set $\{\mathbf{v}_i\}$ spans the entire space S, then there is no loss of information and $\mathbf{x}$ is exactly recoverable from its projections

$$\mathbf{x} = \sum_{i \in \mathcal{Z}} \mathcal{P}_{\mathbf{v}_i}\mathbf{x} \quad \text{[A.4]}$$

Clearly, different choices of basis families give rise to different family of decompositions of the same signal.

When a subset of the projections are used for recovery, or when the transform basis space V is a subspace of the signal space S, one obtains an *approximation* $\mathcal{A}$ of $\mathbf{x}$. The residuals or the unexplained portion of $\mathbf{x}$ is known as the *details*, $\mathcal{D}$. These details can then be treated as projections of $\mathbf{x}$ onto a different subspace W of the signal space S. Thus,

$$\mathbf{x} = \mathcal{A} + \mathcal{D} \quad [A.5]$$

Correspondingly, the coefficient set can be divided into two sets $\{a_j\}$ and $\{d_l\}$ such that

$$\{c_i\} = \{a_j\} \cup \{d_l\}$$

For complex-valued vectors, the projections are real valued, whereas the projection coefficients are complex valued. When the basis space is a continuum, the summation in Eq. (A.2) is replaced by an integral and the coefficient set is also a continuum.

The foregoing concepts are equally valid for functions belonging to Hilbert space. All the interpretations hold good with the inner product defined as

$$\langle f(t), g(t) \rangle = \int_{-\infty}^{\infty} f(t) g^*(t) \mathrm{d}t \quad [A.6]$$

where $g(t)$ is the basis function and the asterisk on the top denotes its complex conjugate.

The Fourier series expansion of a discrete-time periodic signal $x[k]$ constructs a new representation of a *periodic* signal in the space of discrete index complex sinusoids (harmonics) $e^{j\omega_i k}, i \in \mathcal{Z}$. The coefficients are complex valued. On the other hand, the Fourier transform of a finite-energy (2-norm) *a periodic* signal represents the signal in a continuum frequency space spanned by the basis functions $e^{j\omega k}$, $-\pi \leq \omega < \pi$. In both cases, the signal is "transformed" to the space of complex numbers, but the operations are known under different names.

APPENDIX B. PROPERTIES OF THE ESTIMATORS FOR LTI SYSTEMS

In an LTI approximation, the un-modeled nonlinearities are modeled in the subband (scale) indexed by j, by assuming contribution from input and output to be constant over an operating region. It has been observed in Benveniste et al. (1994) that a model with constant parameter at each scale (but that may vary from scale to scale) is an important special case of a general time-varying linear model. Based on the idea, a simplification is possible in the form

$$p_k = k_1\alpha_{jk}, \quad q_k = k_2\alpha_{jk}, \tag{B.1}$$

where intuitively, for one-step ahead prediction, k_1 and k_2 can be seen related to the output autocorrelation and input–output cross-correlation coefficients at lag one. Let us assume that the output measurement is corrupted with stationary, *iid* $N(0, \sigma^2)$ distributed noise. Let superscript s indicates signal component and superscript n indicates noise component in the output measurement. Then substituting Eq. (B.1) in Eq. (3.77), it can be seen that a parameter can be expressed as a sum of a deterministic and a random component.

$$\alpha_{jk} = \overline{\alpha_{jk}} + \Delta\alpha_{jk} \tag{B.2}$$

$$\text{with } \overline{\alpha_{jk}} = \frac{\langle \kappa_k, y_s^s \rangle}{k_1\langle \kappa_k, y^s\rangle + k_2\langle \gamma_k, u\rangle} \text{ and } \Delta\alpha_{jk} = \frac{\langle \kappa_k y_s^n \rangle}{k_1\langle \kappa_k * y^s\rangle + k_2\langle \gamma_k * u\rangle}$$

$$\forall k \in \left\{ I_u : |\langle \gamma_k, u\rangle| \geq \lambda_u \cap I_y : |\langle \kappa_k, y\rangle| \geq \lambda_y \right\}$$

It may be noted that noise in the regressor given by $\langle \kappa_k, y^n\rangle$ is considered to be removed by thresholding and hence the denominator of both the terms on the RHS of Eq. (B.2) are deterministic. Under the assumption that signal and noise components are independent of each other, the uncertainty in the parameter, given by the second term on the RHS of Eq. (B.2), is also zero mean random because

$$E(\Delta\alpha_{jk}) = \frac{E(\langle \kappa_k, y_s^n \rangle)}{(k_1\langle \kappa_k, y^s\rangle + k_2\langle \gamma_k, u\rangle)} = 0 \tag{B.3}$$

where E denotes expectation operator. The variance term of parameter error can be estimated as

$$\hat{P} = E\left((\Delta\alpha_{jk})^2 \right) = \frac{E\left(\langle \kappa_k, y_s^n \rangle^2 \right)}{(k_1\langle \kappa_k, y^s\rangle + k_2\langle \gamma_k, u\rangle)^2} = \frac{\sigma^2}{R^2}, \tag{B.4}$$

where

$$R = (k_1\langle \kappa_k, y^s\rangle + k_2\langle \gamma_k, u\rangle)$$

For a stable system, y^s and u are finite, and hence R is also finite and decides the bound of parameter error.

$$\max(\hat{P}) = \frac{\sigma^2}{\min(\mathrm{R}^2)} = \frac{\sigma^2}{\min\left((k_1\lambda_y)^2, (k_2\lambda_u)^2 \right)} \tag{B.5}$$

It can be seen that the bound of parameter uncertainty can be reduced by increasing thresholds. Hence robustness of the identified model depends on the level of threshold chosen by the designer. At the same time, higher threshold could possibly remove significant signal component, thereby compromising usefulness of the identified model. A trade-off in this regard is necessary to meet the design objective as well as the quality of identification.

APPENDIX C. ALTERNATE PROJECTION ALGORITHM

Following the discussion in Section 7.3, the following theorem can be stated (Mukhopadhyay and Tiwari, 2010).

Theorem 1

Assuming that the noise in the estimate is stationary, iid $N(0,\sigma^2)$ distributed, p_k and q_k are given by $p_k = k_1\alpha_{jk}$, $q_k = k_2\alpha_{jk}$, where k_1 and k_2 are two real-valued constants independent of time, then the first-order estimate of LTI model parameters at scale j based on the consistent output estimate using local error minimization in wavelet domain is given by

$$\hat{\alpha}_j = \frac{1}{M}\sum_{k=I_u \cap I_y}\left[\frac{\langle \kappa_k, y_s\rangle}{k_1\langle \kappa_k, y\rangle + k_2\langle \gamma_k, u\rangle}\right] \quad [C.1]$$

Proof

Substituting $p_k = k_1\alpha_{jk}$, $q_k = k_2\alpha_{jk}$ in Eq. (3.77)

$$\alpha_{jk} = \frac{\langle \kappa_k, y_s\rangle}{k_1\langle \kappa_k, y\rangle + k_2\langle \gamma_k, u\rangle}, \quad \forall k \in \{I_u \cap I_y\} \quad [C.2]$$

Considering size of $I_u \cap I_y$ is M,

$$\hat{\alpha}_j = \frac{1}{M}\sum_{k=I_u \cap I_y}\alpha_{jk} = \frac{1}{M}\sum_{k=I_u \cap I_y}\left[\frac{\langle \kappa_k, y_s\rangle}{k_1\langle \kappa_k, y\rangle + k_2\langle \gamma_k, u\rangle}\right] \quad [C.3]$$

The structure of Eq. (C.1), however, suggests that an iterative scheme can be formulated to find an LTV solution, wherein one iteration, either input or output (not both), is used for prediction such that solution of Eq. (3.77) reduces to estimation of only one parameter. Optimality is reached iteratively by projecting the solution back to time and then again projecting the crude estimate of prediction (in time) forth in transform domain (Mallat and Zhong, 1992). The intermediate solution in transform domain is forced to match significant projection values of the measurement in every iterations such that the prediction is consistent with the measurement (in projections).

REFERENCES

Aadaleesan P, Miglan N, Sharma R, Saha P: Nonlinear system identification using Wiener type Laguerre-Wavelet network model, *Chem Eng Sci* 63:3932–3941, 2008.

Addison P: *The illustrated wavelet transform handbook: introductory theory and applications in science, engineering, medicine and finance,* London, UK, 2002, Institute of Physics.

Akaike H: On the use of a linear model for the identification of feedback systems, *Ann Inst Stat Math* 20:425–439, 1968.

AlZubi S, Islam N, Abbod M: Multiresolution analysis using wavelet, ridgelet, and curvelet transforms for medical image segmentation, *Int J Biomed Imaging* 2011:1–18, 2011.

Auger F, Flandrin P, Lemoine O, Goncalves P: Time-frequency toolbox for MATLAB, 1997. URL http://crttsn.univ-nantes.fr/auger/tftb.html.

Bakshi B: Multiscale analysis and modelling using wavelets, *J Chemom* 13:415–434, 1999.

Bakshi B, Nounou M: Multiscale methods for denoising and compression. In Walczak B, editor: *Wavelets in chemistry, volume 22 of data handling in Science and Technology*, Amsterdam, The Netherlands, 2000, Elsevier Academic Press, pp 119–150.

Bakshi RB, Stephanopoulos G: A multiresolution hierarchial neural network with localized learning, *AIChE J* 39(1):57–81, 1993.

Battle G: A block spin construction of ondelettes. Part I: Lemarie functions, *Commun Math Phys* 110:601–615, 1987.

Benveniste A, Nikoukhah R, Willsky A: Multiscale systems theory, *IEEE Trans Circ Syst I Fund Theor Appl* 41(1):2–15, 1994.

Billings S, Wei H: The wavelet-NARMAX representation: a hybrid model structure combining polynomial models with multiresolution wavelet decompositions, *Int J Syst Sci* 35 (3):137–152, 2005.

Boashash B, editor: *Time-frequency signal analysis*, Australia, 1992, Wiley Halstad Press.

Braatz R, Alkire R, Seebauer E, et al: Perspectives on the design and control of multiscale systems, *J Process Control* 16:193–204, 2006.

Bracewell R: *The Fourier transform and its applications,* ed 3, New York, USA, 1999, Mc-Graw Hill.

Cai C, Harrington P: Different discrete wavelet transforms applied to denoising analytical data, *J Chem Inf Comput Sci* 38:1161–1170, 1998.

Candes E, Donoho D: Ridgelets: a key to higher-dimensional intermittency? *Philos Trans R Soc Lond A: Math Phys Eng Sci* 357(1760):2495–2509, 1999.

Candes E, Donoho D: Curvelets—a surprisingly effective nonadaptive representation for objects with edges. In Cohen A, Rabut C, Schumaker L, editors: *Curves and surface fitting: Saint-Malo*, Nashville, USA, 2000, Vanderbilt University Press, pp 105–120.

Carrier J, Stephanopoulos G: Wavelet-based modulation in control-relevant process identification, *AIChE J* 44(2):341–360, 1998.

Chang C, Fu W, Yi M: Short term load forecasting using wavelet networks, *Eng Intell Syst Electr Eng Commun* 6:217–223, 1998.

Chang X, Qu L: Wavelet estimation of partially linear model, *Comput Stat Data Anal* 47(1): 31–48, 2004.

Chau F, Liang Y-Z, Gao J, Shao X-G: *Chemometrics: from basics to wavelet transform, volume 164 of Analytical Chemistry and its applications,* Hoboken, NJ, USA, 2004, John Wiley & Sons.

Ching P, So H, Wu S: On wavelet denoising and its applications to time delay estimation, *IEEE Trans Signal Process* 47(10):2879–2882, 1999.

Chou C, Willsky A, Benveniste A: Multiscale recursive estimation, data fusion and regularization, *IEEE Trans Autom Control* 39(3):464–478, 1994.

Choudhury M, Thornhill N, Shah S: Modelling valve stiction, *Control Eng Pract* 13:641–658, 2005.

Choudhury M, Shah S, Thornhill N: *Diagnosis of process nonlinearities and valve stiction: data driven approaches,* Berlin, Germany, 2010, Springer.

Christofides P, Daoutidis P: Feedback control of two time-scale non-linear systems, *Int J Control* 63:965–994, 1996.

Chui C, Wang J: On compactly supported spline wavelets and a duality principle, *Trans Am Math Soc* 330(2):903–915, 1992.

Chui CK: *Wavelet: a tutorial in theory and applications,* Boston, 1992, Academic Press.

Claasen T, Mecklenbrauker W: The Wigner distribution—a tool for time-frequency signal analysis—part III: relations with other time-frequency signal transformations, *Philips J Res* 35:372–389, 1980.

Cohen A, Daubechies I, Feauveau J-C: Biorthogonal bases of compactly supported wavelets, *Commun Pure Appl Math* 45:482–560, 1992.

Cohen L: Generalized phase-space distribution functions, *J Math Phys* 7(5):781–786, 1966.

Cohen L: Time-frequency distributions: a review, *Proc IEEE* 77(7):781–786, 1989.

Cohen L: *Time frequency analysis: theory and applications,* Upper Saddle River, New Jersey, USA, 1994, Prentice Hall.

Cole MOT, Keogh PS, Burrows CR, Sahinkaya MN: Wavelet domain control of rotor vibration, *Proc Inst Mech Eng C J Mech Eng Sci* 220(2):167–184, 2006.

Cooley J, Lewis P, Welch P: Historical notes on the fast Fourier transform, *IEEE Trans Audio Electroacoustics* AU-15:76–79, 1967.

Cvetkovic Z, Vetterli M: Discrete time wavelet extrema representation: design and consistent reconstruction, *IEEE Trans Signal Process* 43(3):681–693, 1995.

Daubechies I: Orthogonal bases of compactly supported wavelets, *Commun Pure Appl Math* 41(7):909–996, 1988.

Daubechies I: *Ten lectures in wavelets,* Philadelphia, PA, 1992, Society for Industrial and Applied Mathematics.

Daubechies I, Grossmann A, Meyer Y: Painless nonorthogonal expansions, *J Math Phys* 27:1271–1283, 1986.

Daugmann J: Complete discrete 2-d Gabor transforms by neural networks for image analysis and compression, *IEEE Trans Acoust Speech Signal Process* 36:1169–1179, 1988.

S.U. Department of Statistics: WAVELAB, 2000. http://www-stat.stanford.edu/wavelab.

Desborough L, Miller R: Increasing customer value of industrial control performance monitoring: Honeywell's experience. In *AIChE symposium series no. 326,* vol. 98, 2002, pp 153–186.

Do M, Vetterli M: The contourlet transform: an efficient directional multiresolution image representation, *IEEE Trans Image Process* 14(12):2091–2106, 2005.

Donoho D: De-noising by soft-thresholding, *IEEE Trans Inform Theory* 41(3):613–627, 1995.

Donoho D, Johnstone I: Ideal spatial adaptation by wavelet shrinkage, *Biometrika* 81:425–455, 1994.

Donoho DL, Johnstone LM, Kerkyacharian G, Picard D: Wavelet shrinkage: asymptopia? *J R Stat Soc B* 57:301–309, 1995.

Dorfan Y, Feuer A, Porat B: Model and identification of LPTV systems by wavelets, *Signal Process* 84(8):1285–1297, 2004.

Doroslovacki M, Fan H: Wavelet-based linear system modeling and adaptive filtering, *IEEE Trans Signal Process* 44(5):1156–1165, 1996.

Duffin R, Schaeffer A: A class of nonharmonic Fourier series, *Trans Am Math Soc* 72:341–366, 1952.
Fernandez-Macho J: Wavelet multiple correlation and cross-correlation: a multiscale analysis of eurozone stock markets, *Physica A* 391:1097–1104, 2012.
Fourier J: *The analytical theory of heat (trans: Freeman A),* Cambridge, UK, 1822, Cambridge University Press.
Frano F: *PEM fuel cells: theory and practice,* San Diego, CA, USA, 2005, Elsevier Academic Press.
Gabor D: Theory of communication, *J Inst Electr Eng* 93:429–457, 1946.
Gao R, Yan R: *Wavelets: theory and applications for manufacturing,* New York, USA, 2010, Springer.
Gray F: Pulse code communication. U.S. Patent 2,632,058, March 1953.
Grinsted A, Moore J, Jevrejeva S: Crosswavelet and wavecoherence, 2002. URL http://www.pol.ac.uk/home/research/waveletcoherence/.
Grinsted A, Moore J, Jevrejeva S: Application of the cross wavelet transform and wavelet coherence to geophysical time series, *Nonlinear Process Geophys* 11:561–566, 2004.
Grossmann A, Morlet J: Decomposition of hardy functions into square integrable wavelets of constant shape, *SIAM J Math Anal* 15(4):723–736, 1984.
Haar A: Zur theorie der orthogonalen funktionen-systeme, *Math Ann* 69:331–371, 1910.
Harris T, Seppala C, Desborough L: A review of performance monitoring and assessment techniques for univariate and multivariate control systems, *J Process Control* 9:1–18, 1999.
Harris TJ: Assessment of control loop performance, *Can J Chem Eng* 67:856–861, 1989.
Huang NE, Shen Z, Long SR, et al: The empirical mode decomposition and the Hilbert spectrum for nonlinear and non-stationary time series analysis, *Proc R Soc Lond Ser A Math Phys Eng Sci* 1471-2946454(1971):903–995, 1998. http://dx.doi.org/10.1098/rspa.1998.0193.
Jackson J: Principal components and factor analysis: I. Principal components, *J Qual Technol* 12(4):201–213, 1980.
Jaffard S, Meyer Y, Ryan R: *Wavelets: tools for science and technology,* Philadelphia, PA, USA, 2001, Society for Industrial and Applied Mathematics.
Jelali M: An overview of control performance assessment technology and industrial applications, *Control Eng Pract* 14:441–466, 2005.
Jevrejeva S, Moore J, Grinsted A: Influence of the arctic oscillation and el nin o-southern oscillation (ENSO) on ice conditions in the Baltic sea: the wavelet approach, *J Geophys Res* 108(D21):1–11, 2003.
Juditsky A, Hjalmarsson H, Benveniste A, et al: Nonlinear black-box models in system identification: mathematical foundations, *Automatica* 31(12):1725–1750, 1995.
Kathirmani S, Tangirala A, Saha S, Mukhopadhyay S: Online data compression of MFL signals for pipeline inspection, *NDT & E Int* 0963-869550:1–9, 2012. http://dx.doi.org/10.1016/j.ndteint.2012.04.008. URLhttp://www.sciencedirect.com/science/article/pii/S0963869512000588.
Katic D, Vukobratovic M: Wavelet neural network approach for control of non-contact and contact robotic tasks. In *IEEE symposium on intelligent control, Istanbul, Turkey*, 1997, pp 245–250.
Khalil H: Output feedback control of linear two-time-scale systems, *IEEE Trans Autom Control* AC-32:784–792, 1987.
Kokotovic P, O'Malley R Jr, Sannuti P: Singular perturbation and order reduction in control theory—an overview, *Automatica* 12:123–132, 1976.
Kokotovic P, Khalil HK, O'Reilly J: *Singular perturbations in control: analysis and design,* London, 1986, Academic Press.
Kosanovich K, Moser A, Piovoso M: Poisson wavelet transforms applied to model identification, *J Process Control* 5(4):225–234, 1995.

Krishnan A, Hoo K: A multiscale model predictive control strategy, *Ind Eng Chem Res* 38(5): 1973–1986, 1999.

Lee D: Analysis of phase-locked oscillations in multi-channel single-unit spike activity with wavelet cross-spectrum, *J Neurosci Methods* 115:67–75, 2002.

Lemarie P-G: Ondelettes à localisation exponentielles, *J Math Pures Appl* 67(3):227–236, 1988.

Lio P: Wavelets in bioinformatics and computational biology: state of art and perspectives, *Bioinformatics* 10(1):2–9, 2003.

Ljung L: *System identification—theory for the user,* ed 2, Upper Saddle River, NewJersey, USA, 1999, Prentice Hall PTR.

Lu Z, Sun J, Butts K: Linear programming support vector regression with wavelet kernel: a new approach to nonlinear dynamical systems identification, *Math Comput Simulat* 79:2051–2063, 2009.

Luse D, Khalil H: Frequency domain results for systems with slow and fast dynamics, *IEEE Trans Autom Control* AC-30(12):1171–1178, 1985.

Lutkepohl H: *New introduction to multiple time series analysis,* Berlin, Germany, 2005, Springer.

Ma J, Plonka G: Curvelet transform: a review of recent applications, *IEEE Signal Process Mag* 27(2):118–133, 2010.

Mallat S: Multiresolution approximations and wavelet orthonormal bases of $l^2(r)$, *Trans Am Math Soc* 315(1):69–87, 1989a.

Mallat S: Zero-crossings of wavelet transform, *IEEE Trans Inform Theory* 37(4):1019–1033, 1991.

Mallat S: *A wavelet tour of signal processing,* ed 2, San Diego, CA, USA, 1999, Academic Press.

Mallat S, Zhang Z: Matching pursuits with time-frequency dictionaries, *IEEE Trans Signal Process* 41(12):3397–3415, 1993.

Mallat S, Zhong S: Characterization of signals from multiscale edges, *IEEE Trans PAMI* 14(7): 710–732, 1992.

Mallat SG: A theory for multiresolution signal decomposition: the wavelet representation, *IEEE Trans Pattern Anal Mach Intell* 11:674–693, 1989b.

Maraun D, Kurths J: Cross wavelet analysis: significance testing and pitfalls, *Nonlinear Process Geophys* 11:505–514, 2004.

Mark W: Spectral analysis of the convolution and filtering of non-stationary stochastic processes, *J Sound Vib* 11:19–63, 1970.

Matsuo T, Tadakuma I, Thornhill N: Diagnosis of a unit-wide disturbance caused by saturation in a manipulated variable. In *IEEE advanced process control applications for industry workshop, Vancouver, BC, Canada,* 2004.

Meyer Y: Principe d'incertitude, bases hilberteinnes et algebres d'operateurs. In *Bourbaki seminar,* vol 662, 1985.

Meyer Y: Ondelettes et fonctions splines. In *Seminaire Equations aux Derivees Partielles,* Paris, France, 1986, Ecole Poly-technique.

Meyer Y: *Wavelets and operators. Advanced mathematics,* Cambridge, UK, 1992, Cambridge University Press.

Morlet J, Arens G, Fougean I, Glard D: Wave propagation and sampling theory, *Geophysics* 47:203–236, 1982.

Motard RL, Joseph B: *Wavelet applications in chemical engineering,* MA, USA, 1994, Kluwer Academic Publishers.

Mukhopadhyay S, Tiwari AP: Consistent output estimate with wavelets: an alternative solution of least squares minimization problem for identification of the LZC system of a large PHWR, *Ann Nucl Energy* 37:974–984, 2010.

Mukhopadhyay S, Mahapatra U, Tiwari AP, Tangirala AK: Spline wavelets for system identification. In Kothare M, Tade M, Wouwer AV, Smets I, editors: *DYCOPS 2010: dynamics and control of process systems,* Leuven, Belgium, 2010, IFAC, pp 336–340.

Murtagh F: Wedding the wavelet transform and multivariate data analysis, *J Classification* 15 (2):161–183, 1998.
Ni B, Xiao D, Shah S: Time delay estimation for MIMO dynamical systems—with time-frequency domain analysis, *J Process Control* 20:83–94, 2010.
Nikolaou M, Vuthandam P: Fir model identification: parsimony through kernel compression with wavelets, *AIChE J* 44(1):141–150, 1998.
Ninness B, Gustaffson F: A unifying construction of orthonormal bases for system identification, *IEEE Trans Autom Control* TAC-42(4):515–521, 1997.
Nounou M: Multiscale finite impulse response modeling, *Eng Appl Artif Intel* 19:289–304, 2006.
Nounou M, Bakshi B: On-line multiscale filtering of random and gross errors without process models, *AIChE J* 45(5):1041–1058, 1999.
Nounou M, Nounou H: Multiscale fuzzy system identification, *J Process Control* 15:763–770, 2005.
Nounou M, Nounou H: Improving the prediction and parsimony of ARX models using multiscale estimation, *Appl Soft Comput* 7:711–721, 2007.
Oppenheim A, Schafer R: *Discrete-time signal processing,* Englewood Cliffs, NJ, 1987, Prentice-Hall.
O'Reilly J: Dynamical feedback control for a class of singularly perturbed systems using a full-order observer, *Int J Control* 31:1–10, 1980.
Orfanidis S: *Optimum signal processing,* ed 2, New York, USA, 2007, McGraw Hill.
Paivaa H, Kawakami R, Galvao H: Wavelet-packet identification of dynamic systems in frequency subbands, *Signal Process* 86:2001–2008, 2006.
Palavajjhala S, Motard R, Joseph B: Process identification using discrete wavelet transforms: design of prefilters, *AIChE J* 42(3):777–790, 1995.
Pati Y, Krishnaprasad P: Analysis and synthesis of feedforward neural networks using discrete affine wavelet transformations, *IEEE Trans Neural Netw* 4:73–85, 1992.
Patwardhan SC, Shah SL: From data to diagnosis and control using generalized orthonormal basis filters. Part I: development of state observers, *J Process Control* 15:819–835, 2006.
Patwardhan SC, Manuja S, Narasimhan S, Shah SL: From data to diagnosis and control using generalized orthonormal basis filters, part II: model predictive and fault tolerant control, *J Process Control* 16:157–175, 2006.
Percival D, Walden A: *Wavelet methods for time series analysis, Cambridge series in statistical and probabilistic mechanics,* New York, USA, 2000, Cambridge University Press.
Priestley MB: *Spectral analysis and time series,* London, UK, 1981, Academic Press.
Proakis J, Manolakis D: *Digital signal processing—principles, algorithms and applications,* New Jersey, USA, 2005, Prentice-Hall.
Rafiee J, Rafiee M, Prause N, Schoen M: Wavelet basis functions in biomedical signal processing, *Expert Syst Appl* 38:6190–6201, 2011.
Ramarathnam J, Tangirala AK: On the use of Poisson wavelet transform for system identification, *J Process Control* 19:48–57, 2009.
Reis M: A multiscale empirical modeling framework for system identification, *J Process Control* 19:1546–1557, 2009.
Ricardez-Sandoval L: Current challenges in the design and control of multiscale systems, *Can J Chem Eng* 89:1324–1341, 2011.
Rosas-Orea M, Hernandez-Diaz M, Alarcon-Aquino V, Guerrero-Ojeda L: A comparative simulation study of wavelet-based denoising algorithms. In *15th international conference on electronics, communications and computers*, 2005, IEEE Computer Society, pp 125–130.
Safavi A, Romagnoli J: Application of wavelet-based neural networks to the modelling and optimisation of an experimental distillation column, *Eng Appl Artif Intel* 10(3):301–313, 1997.
Saksena V, O'Reilly J, Kokotovic P: Singular perturbation and time scale methods in control theory: survey 1976–1983, *Automatica* 20(3):273–293, 1984.

Satoa J, Morettina P, Arantes P, Amaro E Jr, : Wavelet based time-varying vector autoregressive modelling, *Comput Stat Data Anal* 51:5847–5866, 2007.
Schuster, A. On lunar and solar periodicities of earthquakes: Proc. Roy. Soc., pp. 455–465, 1897.
Selvanathan S, Tangirala AK: Diagnosis of oscillations due to multiple sources in model-based control loops using wavelet transforms, *IUP J Chem Eng* 1(1):7–21, 2009.
Selvanathan S, Tangirala AK: Diagnosis of poor loop performance due to model-plant mismatch, *Ind Eng Chem Res* 49(9):4210–4229, 2010.
Shan X, Burl J: Continuous wavelet based time-varying system identification, *Signal Process* 91(6):1476–1488, 2011.
Sivalingam S, Hovd M: Use of cross wavelet transform for diagnosis of oscillations due to multiple sources. In Fikar M, Kvasnica M, editors: *18th international conference on process control, Tatranska Lomnica, Slovakia*, 2011, pp 443–451.
Sjöberg J, Zhang Q, Ljung L, et al: Nonlinear black-box modeling in system identification: a unified overview, *Automatica* 31(12):1691–1724, 1995.
Smith M, Barnwell T III : Exact reconstruction for tree structured sub-band coders, *IEEE Trans Acoust Speech Signal Process* 34(3):431–441, 1986.
Smith SW: *Scientist and engineer's guide to digital signal processing,* San Diego, CA, USA, 1997, California Technical Publishing.
Srinivasan B, Tangirala AK: Source separation in systems with correlated sources using NMF, *Digital Signal Process* 20(2):417–432, 2010.
Srinivasan R, Rengaswamy R, Narasimhan S, Miller R: Control loop performance assessment, 2. Hammerstein model approach for stiction diagnosis, *Ind Eng Chem Res* 44(17): 6719–6728, 2005.
Srivastava S, Singh M, Hanmandlu M, Jha A: New fuzzy wavelet neural networks for system identification and control, *Appl Soft Comput* 6:1–17, 2005.
Stein C: Estimation of the mean of a multivariate normal distribution, *Ann Statist* 9 (6):1135–1151, 1981.
Stephanopoulos G, Karsligil O, Dyer M: Multi-scale aspects in model-predictive control, *J Process Control* 10:275–282, 2000.
Strang G, Nguyen T: *Wavelets and filter banks,* Boston, MA, USA, 1996, Wellesley-Cambridge Press.
Sureshbabu N, Farrell J: Wavelet-based system identification for nonlinear control, *IEEE Trans Autom Control* 44(2):412–417, 1999.
Szu H, Telfer B, Kadambe S: Neural network adaptive wavelets for signal representation and classification, *Opt Eng* 31:1907–1916, 1992.
Tabaru T: Dead time measurement methods using wavelet correlation. In *International conference on control, automation and systems, Seoul, Korea*, 2007, pp 2778–2783.
Tabaru T, Shin S: Dead time detection by wavelet transform of cross spectrum data. In *ADCHEM '97: IFAC conference on advanced control of chemical processes*, 1997, pp 311–316.
Takagi T, Sugeno M: Fuzzy identification of systems and its applications to modeling and control, *IEEE Trans Syst Man Cybern* 15:116–132, 1985.
Tangirala AK, Shah S, Thornhill N: PSCMAP: a new tool for plant-wide oscillation detection, *Process Control* 15:931–941, 2005.
Tangirala AK, Kanodia J, Shah SL: Non-negative matrix factorization for detection and diagnosis of plant wide oscillations, *Ind Eng Chem Res* 46:801–817, 2007.
Tewfik AH, Kim M: Correlation structure of the discrete wavelet coefficients of fractional Brownian motion, *IEEE Trans Inform Theory* 38(2):904–909, 1992.
Thao N, Vetterli M: Deterministic analysis of oversampled ad conversion and decoding deterministic analysis of oversampled A/D conversion and decoding improvement based on consistent estimates, *IEEE Trans Signal Process* 42(3):519–531, 1994.

Thornhill N, Horch A: Advances and new directions in plant-wide disturbance detection and diagnosis, *Control Eng Pract* 15(10):1196–1206, 2007.
Thornhill NF, Cox JW, Paulonis MA: Diagnosis of plant-wide oscillation through data-driven analysis and process understanding, *Control Eng Pract* 11:1481–1490, 2003.
Thuillard M: Fuzzy wavenets: an adaptive, multiresolution, neurofuzzy learning scheme. In *EUFIT '99, seventh European congress on intelligent techniques and soft computing, Contrib. cc6-1, CD Proc.*, 1999.
Thuillard M: A review of wavelet networks, wavenets, fuzzy wavenets and their applications, *ESIT 2000* , 2000.
Tiwari A, Bandopadhyay B, Warner H: Spatial control of a large PHWR by piecewise constant periodic output feedback, *IEEE Trans Nucl Sci* 47(2):389–402, 2000.
Torrence C, Compo G: A practical guide to wavelet analysis, *Bull Am Meteorol Soc* 79(1): 61–78, 1998.
Tsatsanis M, Giannakis G: Time-varying system identification and model validation using wavelets, *IEEE Trans Signal Process* 41(12):3512–3523, 1993.
Tzeng S-T: Design of fuzzy wavelet neural networks using the GA approach for function approximation and system identification, *Fuzzy Sets Syst* 161:2585–2596, 2010.
Unser M: Ten good reasons for using spline wavelets. In *SPIEn wavelets applications in signal and image processing,* vol. 3169, 1997, pp 422–431.
Unser M, Aldroubi A: A review of wavelets in biomedical applications, *Proc IEEE* 84(4): 626–638, 1996.
Unser M, Thevenaz P, Aldroubi A: Shift-orthogonal wavelet bases using splines, *IEEE Signal Process Lett* 3(3):85–88, 1996.
Vaidyanathan P: Quadrature mirror filter banks, m-band extensions and perfect reconstruction techniques, *IEEE ASSP Mag* 4(3):4–20, 1987.
Vetterli M: Filter banks allowing perfect reconstruction, *Signal Process* 10(3):219–244, 1986.
Vetterli M: Wavelets, approximations and compression, *IEEE Signal Process Mag* 18(5): 59–73, 2001.
Ville J: Theorie et applications de la signal analytique, *Cables et Transm* 2A(1):61–74, 1948.
Vlachos D: A review of multiscale analysis: examples from systems biology, materials engineering, and other fluid–surface interacting systems, *Adv Chem Eng* 30(1):1–61, 2005.
Wei H, Billings S: Identification of time-varying systems using multiresolution wavelet models, *Int J Syst Sci* 33(15):1217–1228, 2002.
Wei H, Billings S, Zhao Y, Guo L: An adaptive wavelet neural network for spatio-temporal system identification, *Neural Netw* 23:1286–1299, 2010.
Weiss G, Coifman R: Extensions of Hardy spaces and their use in analysis, *Bull Am Math Soc* 83:569–645, 1977.
Wigner E: On the quantum correction for thermodynamic equilibrium, *Phys Rev* 40:749–759, 1932.
Wigner E: Quantum mechanical distribution functions revisited. In Yourgrau W, van der Merwe A, editors: *Perspective in quantum theory*, Boston, MA, USA, 1971, Dover, pp 25–36.
Wold S, Esbensen K, Geladi P: Principal component analysis, *Chem Intell Lab Systt* 2:37–52, 1987.
Xu X, Shi Z, You Q: Identification of linear time-varying systems using a wavelet-based state-space method, *Mech Syst Signal Process* 26:91–103, 2012.
Zekri M, Sadri S, Sheikholeslam F: Adaptive fuzzy wavelet network control design for nonlinear systems, *Fuzzy Sets Syst* 159:2668–2695, 2008.
Zhang Q, Benveniste A: Wavelet networks, *IEEE Trans Neural Netw* 3(6):889–898, 1992.
Zhao H, Bentsman J: Biorthogonal wavelet based identification of fast linear time-varying systems—part I: system representations, *J Dyn Syst Meas Control* 123(4):585–592, 2001a.
Zhao H, Bentsman J: Biorthogonal wavelet based identification of fast linear time-varying systems—part II: algorithms and performance analysis, *J Dyn Syst Meas Control* 123(4): 593–600, 2001b.

CHAPTER FOUR

Multiobjective Optimization Using Genetic Algorithm

Santosh K. Gupta[*,†,1], **Sanjeev Garg**[*]
[*]Department of Chemical Engineering, Indian Institute of Technology, Kanpur, Uttar Pradesh, India
[†]Department of Chemical Engineering, University of Petroleum and Energy Studies (UPES), Dehradun, Uttarakhand, India
[1]Current address: Department of Chemical Engineering, University of Petroleum and Energy Studies (UPES), Dehradun, Uttarakhand, India

Contents

Abstract

Genetic algorithm (GA) is among the more popular evolutionary optimization techniques. Its multiobjective (MO) versions are useful for solving industrial problems that are more meaningful and relevant. Usually, one obtains sets of several equally good (nondominated) optimal solutions for such cases, referred to as Pareto sets. One of the MOGA algorithms is the elitist *n*ondominated *s*orting *g*enetic *a*lgorithm (NSGA-II). Unfortunately, most MOGA codes, including NSGA-II, are quite slow when applied to *real-life* problems and several bio-mimetic adaptations have been developed to improve their rates of convergence. Some of these are described in detail. A few chemical engineering examples involving two or three noncommensurate objective functions are described. These include heat exchanger networks, industrial catalytic reactors for the

Advances in Chemical Engineering, Volume 43
ISSN 0065-2377
http://dx.doi.org/10.1016/B978-0-12-396524-0.00004-0

manufacture of maleic anhydride and phthalic anhydride, industrial third stage polyester reactors, LDPE reactors with multiple injections of initiator, an industrial semibatch nylon-6 reactor, etc. A more compute-intense problem in bio-informatics (clustering of data from cDNA microarray experiments) is also discussed. Some *very* recent bio-mimetic adaptations of NSGA-II that hold promise for greatly improved rates of convergence to the optimal solutions are also presented.

LIST OF SYMBOLS

f_b fixed length of the JG
I_i i-th objective function
l_{chr} length of chromosome
$l_{string,i}$ number of binaries used to represent the i-th decision variable
m number of objective functions
N_{gen} number of generations
$N_{gen,max}$ maximum number of generations
N_p population size
$n_{parameter}$ number of decision variables in GA
N_{seed} random seed
P_{aJG} probability of carrying out the aJG operation
P_{cross} probability of carrying out the crossover operation
$P_{11\cdots1}$ probability for changing all binaries of a selected decision variable to zero
P_{JG} probability of carrying out the JG operation
P_{mJG} probability of carrying out the mJG operation
P_{mut} probability of carrying out the mutation operation
P_{sJG} probability of carrying out the sJG operation
P_{saJG} probability of carrying out the saJG operation
R random number
$\mathbf{X}$, $\mathbf{x}$ vector of decision variables, X_i or x_i

1. INTRODUCTION

1.1. Overview

Optimization techniques have long been applied to problems of industrial importance. Several excellent texts (Beveridge and Schechter, 1970; Bryson and Ho, 1969; Deb, 1995; Edgar et al., 2001; Gill et al., 1981; Lapidus and Luus, 1967; Ray and Szekely, 1973; Reklaitis et al., 1983) describe the various "traditional" methods with examples. These usually involve the minimization of a *single*-objective function, $I(\mathbf{x})$, or the maximization of $F(\mathbf{x})$, with bounds on the several decision (design or control) variables, $\mathbf{x}\left(\equiv \left[x_1, x_2, \ldots, x_{n_{parameter}}\right]^T\right)$. A *unique* optimal solution is often obtained. A simple example involving two ($n_{parameter} = 2$) decision variables is given by

$$\begin{aligned}&\text{Max } F(\mathbf{x}) \text{ or Min } I(\mathbf{x})\\&\text{subject to (s.t.)}:\\&\text{bounds on } \mathbf{x}: x_i^{\mathrm{L}} \le x_i \le x_i^{\mathrm{U}};\quad i=1,2\end{aligned} \qquad [4.1]$$

Most real-world engineering problems, however, require the simultaneous optimization (maximization *or* minimization) of *several* objectives that cannot be compared easily with each other, that is, are noncommensurate. These are referred to as multiobjective optimization, MOO, problems. For example, the satiation of the palate by apples and the satiation by oranges involve two separate, noncommensurate objectives. These cannot be combined into a single, meaningful scalar objective function by adding the two with weighting factors, something that was done routinely over 25 years ago. A simple, two-objective example (any combination of maximization and minimization) involving two decision variables ($n_{\text{parameter}} = 2$) is described by

$$\begin{aligned}&\text{Min } I_1(\mathbf{x}) \text{ or Max } F_1(\mathbf{x})\\&\text{Min } I_2(\mathbf{x}) \text{ or Max } F_2(\mathbf{x})\\&\text{s.t.}:\\&\text{bounds on } \mathbf{x}: x_i^{\mathrm{L}} \le x_i \le x_i^{\mathrm{U}};\quad i=1,2\end{aligned} \qquad [4.2]$$

The objective function now becomes a vector. A chemical engineering example (Sankararao and Gupta, 2007a) is the *maximization* of the yield of gasoline (the desired product) from the riser reactor of a fluidized-bed catalytic cracking unit (FCCU, see Fig. 4.1), while simultaneously *minimizing* the percentage of CO in the flue gas released from the regenerator [it may be mentioned that a problem involving the minimization of I_i can be solved in terms of the maximization of F_i, using a common but not unique transformation: Min $I_i \rightarrow$ Max F_i $\{\equiv 1/(1+I_i)\}$]. Often, instead of obtaining a single (unique) optimal solution as in the single-objective problem, we obtain a set of several equally good (nondominated) optimal solutions, called a Pareto front. An example of the Pareto set for the two-objective optimization problem for the FCCU is shown in Fig. 4.2. It is clear that point B is better (superior; higher) in terms of the gasoline yield but worse (inferior; higher) in terms of the CO emitted than A. Points A and B are referred to as *nondominated* points since neither is superior to (dominates over) the other. The *entire* set of points shown by diamonds in Fig. 4.2 has this characteristic, and hence represents a Pareto front. Point C, however, does not have this property. In fact, C is inferior to point A with respect to both the objective functions (lower conversion and higher CO). Point A is said to dominate over point C.

Each point in Fig. 4.2 is associated with a set of values of the decision variables, $\mathbf{x}$. At the present moment, we can provide a design engineer,

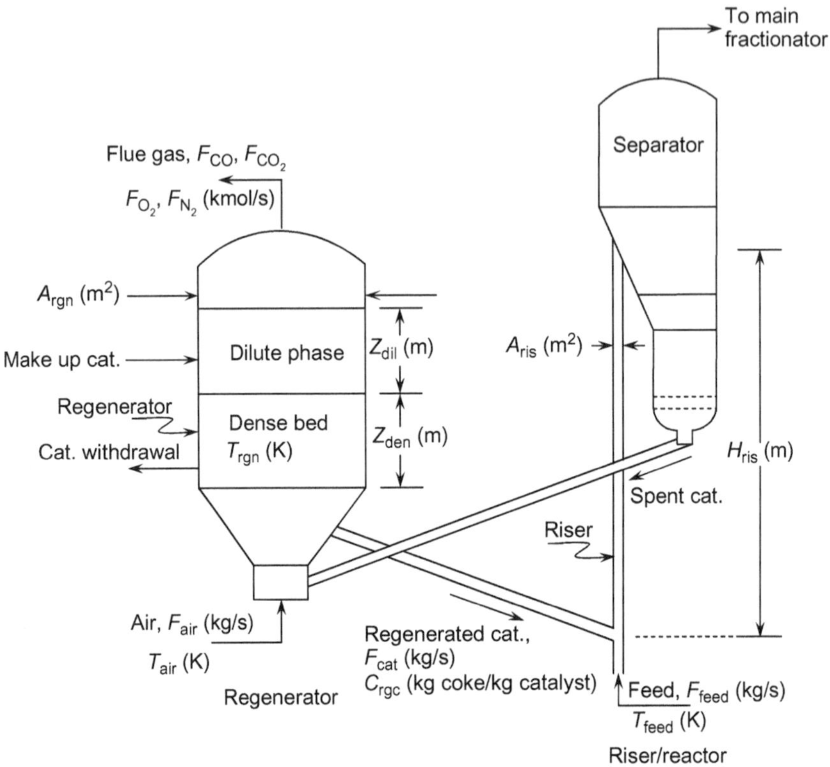

Figure 4.1 Schematic of a fluidized-bed catalytic cracking unit (FCCU).

called a decision maker, with a Pareto set of optimal solutions from among which he/she can select a suitable operating point (called the preferred solution). Often, this decision involves some amount of nonquantifiable intuition. Work along the lines of making this second step easier is a focus of current research. In Fig. 4.2, it is easy to select the preferred solution. A point slightly to the left of *D* would appear to be the best, as beyond this point there is little improvement/increase in the gasoline yield, but a significant worsening of the CO emission.

1.2. The ε-constraint method for obtaining Pareto fronts

An early approach to solve such two- (or multi-) objective optimization problems is the ε-constraint method (Haimes, 1977; Haimes and Hall, 1974; Wajge and Gupta, 1994). For a two-objective optimization problem (Eq. 4.2) one solves a more highly constrained single-objective problem (from here on, we use minimization of all the objective functions, I_i, for

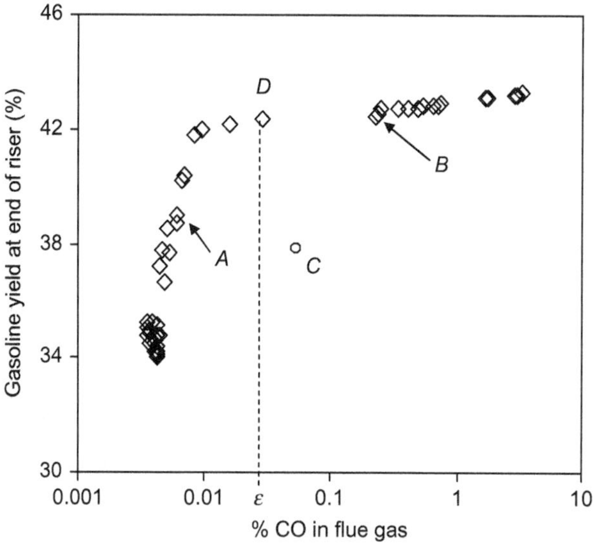

Figure 4.2 The Pareto set obtained for the FCCU problem. An additional point, C, is also indicated. *Adapted from Sankararao and Gupta (2007a).*

illustration; it is easy to replace any of these by Max F_i, if any of the objective functions is to be maximized)

$$\begin{aligned}&\text{Min } I_1(\mathbf{x}) \ [\text{or, Min } I_2(\mathbf{x})]\\&\text{s.t.:}\\&x_i^{\text{L}} \le x_i \le x_i^{\text{U}};\quad i=1,2\\&I_2(\mathbf{x})[\text{or } I_1(\mathbf{x})]=\varepsilon\end{aligned} \tag{4.3}$$

where ε is a specified constant. Figure 4.2 shows one such choice of ε. Any optimization technique, for example, Pontryagin's maximum/minimum principle (Beveridge and Schechter, 1970; Bryson and Ho, 1969; Edgar et al., 2001; Ray and Szekely, 1973), sequential quadratic programming (SQP), GA, SA, etc., may be used for solving Eq. (4.3). The ε-constraint method finally gives point D (in Fig. 4.2) as the final solution of Eq. (4.3). Solving Eq. (4.3) for several choices of ε will give the entire Pareto set. If the MOO problem involves more than two- (say, p) objective functions, one constrains *any* $p-1$ objectives as

$$I_i(\mathbf{x})=\varepsilon_i;\quad i=1,2,\ldots,p-1 \tag{4.4}$$

and solves the resulting single-objective problem. Wajge and Gupta (1994) have used Pontryagin's principle to solve a two-objective optimization problem for a nonvaporizing industrial nylon-6 reactor using this method.

2. BINARY-CODED GENETIC ALGORITHM FOR SINGLE-OBJECTIVE PROBLEMS

A very popular and robust technique for solving optimization problems is genetic algorithm (GA). The binary-coded single-objective version, namely, simple GA (SGA) (Coello Coello et al., 2007; Deb, 1995, 2001; Goldberg, 1989; Holland, 1975) is described first, followed by its extensions and bio-mimetic adaptations for multiobjective (MO) cases.

The single-objective optimization problem we have chosen for illustration is given by Eq. (4.1). We first generate, using a *series* of several random numbers, R_i, a population of N_p solutions (called parent chromosomes or strings), each representing a set of $n_{parameter}$ decision variables. Each decision variable is represented in terms of l_{string} *binary* numbers. Thus, there are $n_{chr} \equiv l_{string} \times n_{parameter}$ binaries in any chromosome and there are a total of $n_{chr}N_p$ binary numbers in the entire population. An easy way to generate such a set of binary numbers in the population randomly is to assign an "event" to a "range" of the generated random number. For example, we generate a sequence of $n_{chr}N_p$ random numbers (usually $0 \leq R_i \leq 1$) and use, say, zero (the "event") if R_i comes out in the range $0 \leq R_i \leq 0.49999$, while we select unity if $0.5 \leq R_i \leq 1$. An example of two chromosomes generated using a random number generating code and with $l_{string} = 4$, $n_{parameter} = 2$, is

	S_3	S_2	S_1	S_0	S_3	S_2	S_1	S_0	
1st chromosome:	1	0	1	0	0	1	1	1	
2nd chromosome:	1	1	0	1	0	1	0	1	[4.5]
	Decision variable (substring) 1				Decision variable (substring) 2				

In Eq. (4.5), S_0, S_1, S_2, and S_3 denote the binaries in any substring at the zeroth, first, second, and third positions (from the right end), respectively.

We now map these binaries representing the decision variables into real numbers, ensuring that the bounds are satisfied. The domain, $[x_i^L, x_i^U]$, for decision variable, x_i, is divided into $(2^{l_{string}} - 1)$ [=15 in the present example with $l_{string} = 4$] *equi-spaced* intervals and all the 16 possible binary numbers assigned sequentially. In Fig. 4.3, the lower bound, x_i^L, for decision variable, x_i, is assigned to the "all 0" substring, (0 0 0 0), while the upper limit, x_i^U, to the "all 1" substring, (1 1 1 1). The other binary substrings are assigned *sequentially* between the bounds of x_i, (see Fig. 4.3). It is easy to map (decode) a binary substring into a *real* value using

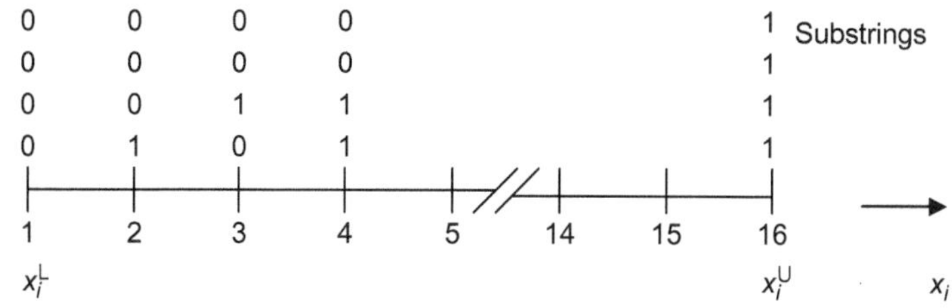

Figure 4.3 Bounds and mapping of binary substrings, $l_{\text{string}}=4$.

$$x_i = x_i^L + \frac{x_i^U - x_i^L}{2^{l_{\text{string}}} - 1} \times \left(\sum_{i=0}^{l_{\text{string}}-1} 2^i S_i \right) \qquad [4.6]$$

The larger the l_{string}, the more accurate is the search. The mapped real values of each of the two decision variables in Eq. (4.5) are used in a model to evaluate the value of the objective function, $I(\mathbf{x}_j)$. This is done for each of the j chromosomes, $j=1, 2, \ldots, N_p$, in the population.

The N_p feasible solutions (parent chromosomes; generation number, $N_{\text{gen}}=1$), each associated with an objective function, need to be improved to give N_p daughter chromosomes (which will be the new parents in the next generation, $N_{\text{gen}}=2$) by mimicking natural genetics. This is done using a three-step procedure. The first step is referred to as *copying* or *reproduction*. We make N_p copies of the parent chromosomes at a new location, called the "gene pool." This is done randomly using another sequence of random numbers, R (the subscript on R is being dropped). The *tournament selection* procedure can be used (other techniques are available, Deb, 2001; Coello Coello et al., 2007). If $N_p=100$, $0 \le R \le 0.01$ (the range of R) is assigned to chromosome number 1 (the event), $0.01 \le R \le 0.02$ is assigned to chromosome number 2, etc. Two random numbers are generated sequentially, and two corresponding chromosomes are selected and compared. The better of these two chromosomes [in terms of the values of the objective functions, $I(\mathbf{x}_j)$] is copied in the gene pool (without deleting either of the two from the pool of the parent chromosomes). This procedure is repeated N_p times. Clearly, chromosomes having better values of I are selected more frequently in the gene pool. Due to the randomness associated with this copying procedure, there are chances that some poor chromosomes also get copied (survive). This helps maintain diversity of the gene pool (two morons can produce a genius!). Also, multiple copies of the superior parent chromosomes can be present in the gene pool.

The *crossover* operation is now carried out on the N_p copies of the parent chromosomes in the gene pool. This is similar to what happens in biology.

The chromosomes in the gene pool are assigned a number from 1 to N_p. We first select two strings in the gene pool, randomly, again, using an appropriate assignment of R to the N_p members in the gene pool. We then check if we need to carry out crossover (as described later) on this pair, using a specified value of the crossover probability, P_{cross}. A random number in the range [0, 1] is generated for the selected pair. Crossover is performed (as described later) if this number happens to lie between 0 and P_{cross}. If the random number lies in $[P_{cross}, 1]$, we copy the pair without carrying out crossover. This procedure is repeated $N_p/2$ times to give N_p daughter chromosomes, with $100(1-P_{cross})\%$ of these being copies (as of now) of the parents. This helps in preserving some of the *elite* members of the parent population in the next generation (an additional, more powerful version of elitism is described later). It may be noted that the chromosomes in the gene pool remain there and could possibly be selected again.

Crossover involves selection of a location (crossover site) in the string, randomly, and swapping the two strings at this site, as shown below:

$$\begin{array}{ccc} 1001\ \ 1 \vdots 100 & & 1001\ \ 1 \vdots 101 \\ 1011\ \ 0 \vdots 101 & \Rightarrow & 1011\ \ 0 \vdots 100 \\ \text{Parent chromosomes} & & \text{Daughter chromosomes} \end{array} \quad [4.7]$$

In the above example, there are seven possible *internal* crossover sites. We assign ranges, $0 \le R \le 1/7; 1/7 \le R \le 2/7; \ldots; 6/7 \le R \le 1$ of a (another set of) random number to each of these seven crossover sites and carry out this operation as shown in Eq. (4.7).

If we somehow have a population in which all parent chromosomes happen to have a 0 in, say, the first location, for example, in

$$\begin{array}{c} 0101\ \ 1\ \ 100 \\ 0001\ \ 1\ \ 101 \\ 0011\ \ 0\ \ 101 \end{array} \quad [4.8]$$

it will be impossible to get a 1 at this position using crossover. If the optimal solution happens to have a 1 at the first position, we will not be able to obtain it. A similar problem exists for all locations in the chromosomes. This drawback is overcome through another bio-mimicked mutation operation. This is carried out after the crossover is completed. In this, *every* $N_p N_{chr}$ binary in the daughter population (including the ones that are copied without crossover from the parents), are changed from 0 to 1 or *vice versa* with a small mutation probability, P_{mut}. That is, if a new (another set) random number corresponding to any binary lies between 0 to P_{mut}, mutation is performed. Too large a value of P_{mut} leads to oscillations of the solutions.

These three operations complete the generation of N_p daughter chromosomes. The process is repeated (till convergence or for a prespecified number, $N_{gen,max}$, of generations) with the daughter chromosomes becoming the parents in the next generation.

The crossover and mutation operations may create inferior strings but we expect them to be eliminated over successive generations by the reproduction operator (the Darwinian principle of survival of the fittest). Since SGA works probabilistically with several solutions simultaneously, we can get multiple optimal solutions, if present. For the same reason, SGA does not (normally) get stuck in the vicinity of local optimal solutions, and so is quite robust.

Optimization problems often involve additional (than bounds) constraints of the kind, $g_i(\mathbf{x}) \leq 0; i=1, 2, \ldots, p$. The penalty function approach (Beveridge and Schechter, 1970; Deb, 1995; Edgar et al., 2001) is used to take care of these (Deb, 2001 suggests an alternate methodology to penalize chromosomes violating constraints). In the penalty function approach, the constraints are added to I (for a minimization problem) in a weighted form when the constraint is violated and the modified objective function minimized (for a maximization problem, an appropriate penalty function is subtracted when the constraint is violated). The following example (Deb, 1995) illustrates the procedure

$$\begin{aligned}
&\text{Min } I(x_1,x_2) = \left[\left(x_1^2+x_2-11\right)^2+\left(x_1+x_2^2-7\right)^2\right] && [a]\\
&\text{s.t.:}\\
&\text{constraint: } (x_1-5)^2+x_2^2-26 \geq 0 && [b]\\
&\text{bounds: } 0 \leq x_1 \leq 5;\quad 0 \leq x_2 \leq 5 && [c]
\end{aligned} \qquad [4.9]$$

The modified objective function is written as (as our code maximizes the objective function):

$$\text{Max } F(x_1,x_2) = \frac{1}{1+\left[\left(x_1^2+x_2-11\right)^2+\left(x_1+x_2^2-7\right)^2\right]} - w_1\left[(x_1-5)^2+x_2^2-26\right]^2$$

$$\text{s.t.: bounds: } 0 \leq x_1 \leq 5;\ 0 \leq x_2 \leq 5 \qquad [4.10]$$

In Eq. (4.10), w_1 is taken to be a large positive number (depending on the value of the original objective function) in case the constraint is violated; else it is assigned a value of zero. It is clear that when the constraint is violated, the value of the modified objective function in Eq. (4.10) decreases, thus favoring the elimination of that chromosome over the next few generations. Equality constraints can be handled in a similar manner. The results for this problem are shown in Fig. 4.4 for two values of N_{gen}. Figure 4.4 shows that most of the $N_p = 100$ solutions lie around the optimal (constrained) point, $\mathbf{x} = (0.829, 2.933)^T$, at $N_{gen} = 7$, but all the hundred solutions are identical (converged) and lie at the optimal point at $N_{gen} = 16$. It must be cautioned that real-life MOO problems will not converge to the optimal solution so early, and one has to try out several values of the computational parameters, P_{cross}, P_{mut}, N_{genmax}, N_p, l_{str}, w_1, etc. For computationally intensive problems that are common in chemical engineering, P_{cross} ranges typically from 0.95 to 0.99, P_{mut} from 0.005 to 0.05, N_{genmax} usually ranges from 100 to 200 (but higher values of the order of 2,500,000 have also been used in array informatics problems), N_p is typically 100 (but larger values of 1000 have also been used for some problems), l_{str} ranges from 32 to 64, and w_1 is typically 10^5–10^6. Table 4.1 gives typical values of these computational parameters for some simple benchmark (test) problems discussed later. In fact, one may also have to use several problem-specific "tricks" to converge to the optimal solution! These are described for individual problems later.

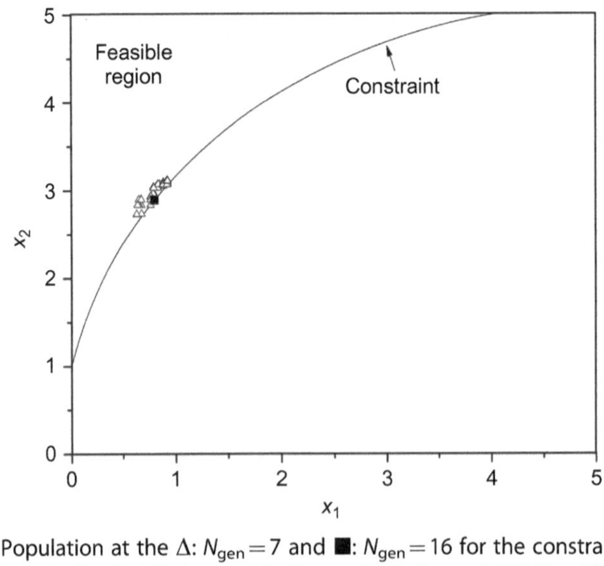

Figure 4.4 Population at the Δ: $N_{gen} = 7$ and ■: $N_{gen} = 16$ for the constrained optimization problem in Eq. (4.10). $l_{string} = 16$, $P_{cross} = 0.95$, $P_{mut} = 0.03125 = 1/32$, $N_p = 100$, and $w_1 = 10^5$.

Table 4.1 Computational parameters for NSGA-II-aJG and NSGA-II-JG for Problems 1–3 (Agarwal and Gupta, 2008a)

Parameter	Problem 1 (ZDT2)	Problem 2 (ZDT3)	Problem 3 (ZDT4)
N_p	100	100	100
$N_{gen,max}$	1000	1000	1000
N_{seed}[a]	0.88876	0.88876	0.88876
l_{chr}	900	900	300
P_{cross}	0.9	0.9	0.9
P_{mut}	0.01	0.01	0.01
P_{JG}	0.40	0.50	0.50
P_{aJG}	0.40	0.30	0.50
$f_{b,aJG}$	25	25	25

[a]N_{seed} is a parameter required by the code generating random numbers (and controls their sequence).

SGA has several advantages over traditional techniques. This technique is better than calculus-based methods (both direct and indirect methods) that may get stuck at local optima, and that may miss the global optimum. This technique does not need derivatives, either.

3. MO ELITIST NONDOMINATED SORTING GA, NSGA-II

SGA has been extended by several workers (Coello Coello et al., 2007; Deb, 2001; Jaimes and Coello Coello, 2009) to give Pareto-optimal solutions for MOO problems. Of these, the *n*ondominated *s*orting *g*enetic *a*lgorithm (NSGA-II; Deb, 2001; Deb et al., 2002) has become quite popular. This algorithm and its predecessor, NSGA-I, uses the concept of elitism, that is, the better (elite) parents as well as the better daughters are used as parents in the next generation, something that does not happen in biology. These algorithms have been used to solve several MOO problems in chemical engineering (Bhaskar et al., 2000a). The binary-coded NSGA-II is now described in detail using the unconstrained two-objective Eq. (4.2) as an example. Constraints of the kind given in Eq. (4.9b) can be handled using the penalty function approach (as in Eq. 4.10).

Figure 4.5 gives a simplified flowchart of NSGA-II. N_p parent binary chromosomes, C_i; $i=1, 2, \ldots, N_p$, are generated randomly as in SGA (see box P in Fig. 4.5). These are mapped between the bounds of $\mathbf{x}$, and

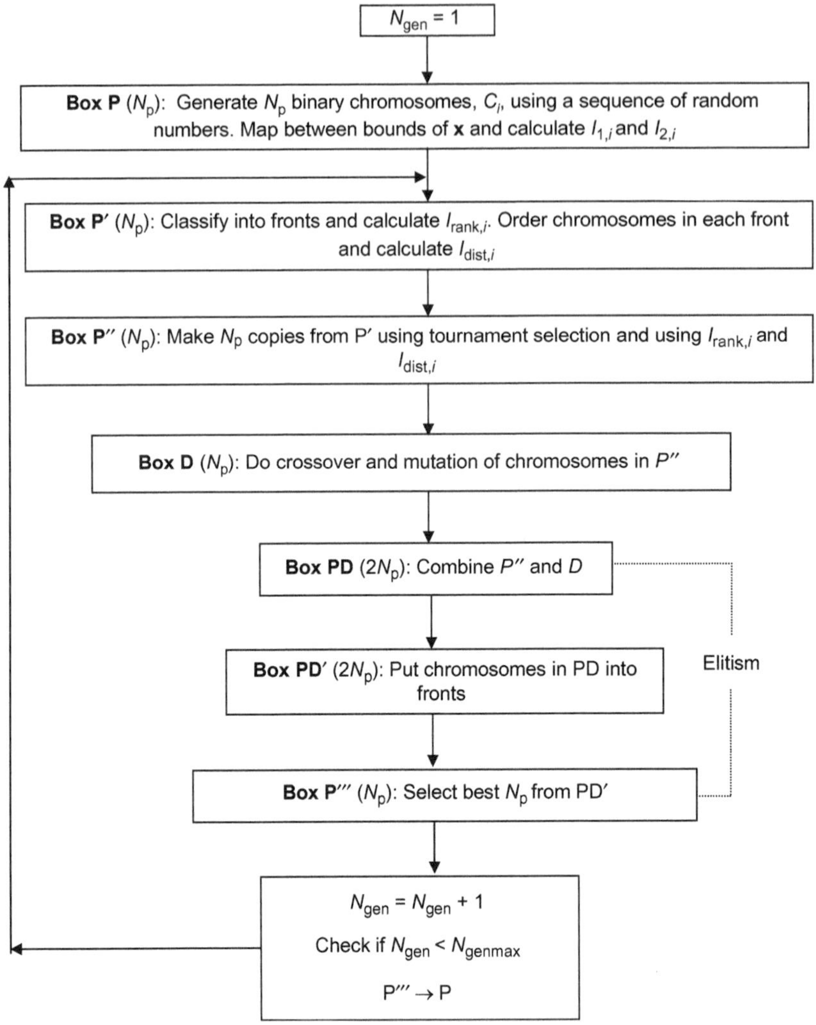

Figure 4.5 Flowchart for NSGA-II for a two-objective optimization problem.

the values of both $I_{1,i}(\mathbf{x})$ and $I_{2,i}(\mathbf{x})$; $i = 1, 2, \ldots, N_p$, for each is obtained. We select the best nondominated subset of chromosomes from these N_p, as described next. The *first* chromosome, C_1, is copied in box P′ having N_p vacant positions (transferred, deleting it from P; see Fig. 4.5). Then the next chromosome, C_2, is transferred temporarily to this box and the two compared using $I_{1,1}$, $I_{2,1}$ with $I_{1,2}$, $I_{2,2}$. If C_2 dominates over C_1 (i.e., both $I_{1,2}$ and $I_{2,2}$ of C_2 are better than the two objective functions, $I_{1,1}$, $I_{2,1}$, of C_1) C_1 is sent back *to its place* in box P. If C_1 dominates over C_2, C_2 is returned *to its place* in P. In other words, the inferior point is removed from P′ and put back into P at its old position. If C_1 and C_2 are nondominated, both are kept in P′. This procedure is repeated with the next chromosome in box P, that is, C_3. At any stage (when C_i is transferred to P′), it is compared with *each* of the existing members in P′, one by one, and the chromosomes that are dominated over (including C_i) are sent back to their locations in P. This is done till all N_p members in P have been so explored. At the end, a subset of nondominated chromosomes are left in P′. We say that these comprise the first (and best) *front*, and assign *all* of these chromosomes a rank of 1 (i.e., $I_{rank,i}=1$ for all chromosomes in front 1). We now "close" this subbox in P′ and generate further fronts (with $I_{rank,i}=2, 3, \ldots$) which are nondominated *within themselves*, but are worse than those in the previous fronts (the comparison in any later subbox is only with the chromosomes present in *that* subbox). This is continued till all N_p chromosomes are sorted (and transferred to P′) using the concept of nondominance. This gives the algorithm its name. It is obvious that *all* the chromosomes in front 1 are the best and are equally good, followed by those in fronts 2, 3, . . .

The Pareto set finally obtained should not only have nondominated members, but have a good *spread* over the domain of **x** or **I**. To get this, we try to de-emphasize (kill slowly) solutions that are closely spaced. This is done by assigning a *crowding distance*, $I_{dist,i}$, to each chromosome, C_i, in P′. For members of any front, we rearrange its chromosomes in order of increasing values of $I_{1,i}$ (or $I_{2,i}$), and find the *size* (sum of all the sides) of the largest cuboid formed by its nearest neighbors in the **I** space. The lower the value of $I_{dist,i}$, the more crowded is the chromosome, C_i. Boundary chromosomes are assigned (arbitrarily) high values of $I_{dist,i}$ (this is somewhat hidden in the available codes and one needs to be careful), so as to prevent their being killed.

The chromosomes in P′ are now copied in a gene pool (box P″) using tournament selection (clearly, if we look at two chromosomes, i and j, in P′ selected randomly, C_i is better than C_j if $I_{rank,i}<I_{rank,j}$. If, however, $I_{rank,i}=I_{rank,j}$, then C_i is better than C_j if $I_{dist,i}>I_{dist,j}$). Crossover and mutation are now carried out on the chromosomes in P″ and the N_p daughter chromosomes stored in D.

The N_p (better) parents (in box P″) and the N_p daughters (in D) are copied into a new box, PD. These $2N_p$ chromosomes are reclassified into fronts (in box PD′), using the concept of domination. The *best* N_p chromosomes are taken from these and put into box P‴, front-by-front. In case only a few members are needed from the *last* front in PD′ to fill up P‴ (as we have to choose N_p from $2N_p$), the least crowded chromosomes from the last front are selected. It is clear that this procedure, called elitism (Deb, 2001), collects the best members from the parents and the daughters. The concept of elitism does not occur in actual genetics. However, it improves the performance of the algorithm significantly.

This completes one generation (N_{gen} is increased by one). The members in P‴ are the parents in the next generation unless appropriate stopping conditions are satisfied, the most common being N_{gen} exceeding the maximum specified number of generations, N_{genmax}.

4. BIO-MIMETIC JUMPING GENE (TRANSPOSON; STRYER, 2000) ADAPTATIONS

NSGA-II as well as similar GA-based MOO algorithms require large amounts of computational (CPU) time for *real-life* MOO problems. Any adaptation to speed up the solution procedure is, thus, desirable. An attempt has been made along this direction by Agarwal and Gupta (2008a,b), Bhat et al. (2006), Bhat (2007), Chan et al. (2005a,b); Guria et al. (2005a), Kasat and Gupta (2003), Man et al. (2004), Ripon et al. (2007), and Simoes and Costa (1999a,b) to improve NSGA-II using the concept of jumping genes (JG; or transposons, predicted by McClintock, 1987) in biology. In fact, the JG adaptation of Kasat and Gupta (2003) has been mentioned by Jaimes and Coello Coello (2009) to be one of the four "most significant multiobjective evolutionary algorithms (MOEAs) that originated in the Chemical Engineering literature."

In biology, JG is a DNA of about a thousand base-pairs that can *jump* in and out of chromosomes. Initially, the idea of JG met with considerable cynicism, but as experimental techniques developed over time, scientists succeeded (in the late 1960s) in isolating JGs from *E. coli* (McKlintock received the 1983 Nobel prize in medicine for her discovery of JGs). In the 1970s, the role of transposons in transferring bacterial resistance to antibodies became understood. It was found that transposons also generated genetic variations (diversity) in natural populations, and that these could confer properties such as drug resistance and toxigenicity, and, under appropriate conditions, offer advantages in terms of survival. The concept of JG has been bio-mimicked to give several JG adaptations of NSGA-II. One of these, namely, NSGA-II-JG, is now described (Kasat and Gupta, 2003; Ramteke and Gupta, 2009a).

Kasat and Gupta (2003) found that the binary-coded NSGA-II can be improved significantly by replacing segments of binaries (genes) by randomly generated JG (see Fig. 4.6). A chromosome in box D (Fig. 4.5), after crossover and mutation, is checked to see if the JG adaptation is to be carried out on it, using a probability, P_{JG} (i.e., if a random number is in the range, $0 \leq R \leq P_{JG}$, the JG operation is carried out on that chromosome). If so, two locations, p and q (both integers), are identified randomly on it, and the binary substring in-between these points is replaced with a *newly* (randomly) generated string (*rs* in Fig. 4.6) of the same length (using the same procedure as for generating the parent chromosomes in $N_{gen} = 1$). Only a single transposon is assumed to be present in any selected chromosome. This is done to keep the algorithm, NSGA-II-JG, simple. The replacement procedure involves a macro–macro-mutation, provides higher genetic diversity, and has usually been found to be superior to NSGA-II. Values of P_{JG} of about 0.4–0.6 seem to work well.

More recently, another, even more powerful, adaptation of JG, NSGA-II-aJG, has been developed by Bhat et al. (2006) and Bhat (2007) (also used by Khosla et al., 2007; Guria et al., 2005a). In this, a probability, P_{aJG}, is used to see if a chromosome in D (Fig. 4.5) after crossover and mutation, is to be modified. If yes, a single site, only, is identified (randomly) in it. The other end of the JG is selected at a (user specified) constant distance (an integer), f_b, beyond it. The substring of binaries in-between these sites is replaced with a *newly* (randomly) generated binary string having the same length. The improved performance of this adaptation probably stems from the introduction of yet another computational parameter, f_b. Values of f_b equal to $l_{str}/8$ seem to be

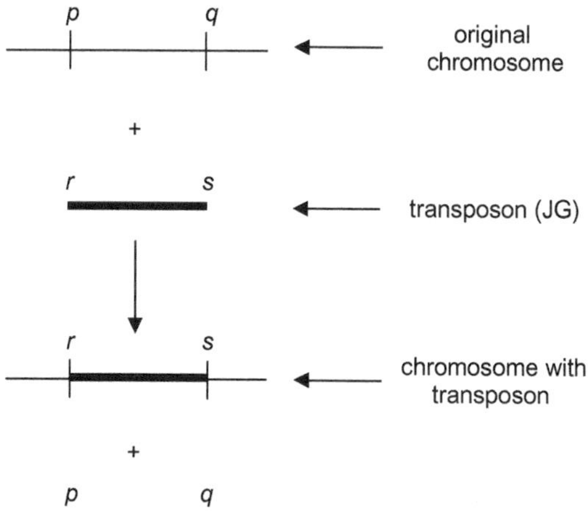

Figure 4.6 The replacement by a JG in a chromosome.

useful. It has been our experience that NSGA-II-aJG works better for several chemical engineering problems than does NSGA-II-JG.

Several bio-mimetic adaptations of JG have been developed for network problems. Guria et al. (2005b) developed the modified jumping gene (mJG) operator for froth flotation circuits, while Agarwal and Gupta (2008a,b) developed the binary-coded NSGA-II-saJG and NSGA-II-sJG for the MOO of heat exchanger networks (HENs), with $f_b = l_{string}$ and the starting location of the JG either being *anywhere* in the chromosome (saJG), or only at the beginning of binaries describing any decision variable (sJG). In the latter case, it is clear that only one decision variable is replaced. Speeding up of the real-coded NSGA-II (discussed later) using the JG adaptation has been observed by Ripon et al. (2007). Hence, the JG operator is a useful adaptation for NSGA-II for the solution of complex MOO problems. Indeed, Sharma et al. (2013) have compared the several JG adaptations on benchmark problems described later.

It is observed that for array informatics applications (grouping genes into clusters with similar gene expressions from microarray experiments for observing differential expression and functional annotations, etc., and gene network analyses, as described below), NSGA-II with the JG operator fails to converge to the average cluster profiles. This is attributed to the dimensionality of the data and the subsequent divergence of GAs due to its probabilistic nature.

We start with a short discussion of cDNA microarray experiments. cDNA microarray technology has been a major revolution in genomics. Presently, microarrays are widely used in laboratories throughout the world to measure the expression levels of tens of thousands of genes *simultaneously* on a single chip. Microarrays are ordered sets (spots) of DNA molecules of known sequences usually representing a gene. Two DNA strands (or one DNA strand and the other an mRNA strand) will hybridize (form complementary base-pair bonds) with each other, regardless of whether they originated from a single source or from two different sources, as long as their base-pair sequences match according to the complementary base-pairing rules. This tendency of complementary DNA strands to hybridize is used in microarrays. The process involves hybridization of unknown gene sequences (samples), which are mobile, over known gene sequences, immobilized over the surface of the chip. The immobilized phase is called as the *probe*, while the mobile phase is termed as the *target*. One of two fluorescent (fluor) tags (*cy3* or G, and *cy5* or *R*) is attached to the probe and the other to the target to quantify their expressions. Complementary base-pairing rules are used to match the unknown sequences with the known sequences after hybridization. The microarray is scanned to determine how much of each probe is bound at each of the several spots. The microarray is placed in a dark room and then stimulated with lasers. The emitted light is captured by a detector which records the fluorescence intensity of the light at each spot. Each of the two fluors used has a characteristic excitation wavelength that will cause the tags to fluoresce. The intensity of the light captured is a measure of the gene expression under the experimental conditions. It is related to the biological function of the genes and their activity. From these values of the intensities, a ratio is calculated which is then interpreted for biological activity. If the R intensity is greater than of the G, then the spot will appear red and that gene is said to be overexpressed or upregulated. If the G intensity is greater than that of R, the spot will appear green and that gene is then underexpressed or downregulated. If the intensities of both R and G are equal than we get a yellow spot, which means that the gene is equally expressed. A black spot on the microarray indicates that at that position no hybridization has occurred.

After a series of image processing steps and data normalization procedures, the microarray data obtained is in the form of an $n \times m$ matrix, where n represents the number of genes (typically, in thousands) and m represents the number of experiments or time-series points (typically, less than a hundred). This data is analyzed for useful biological information. The measured amount of upregulation and underregulation is sorted out using various computational algorithms including GA. Genes are grouped in the form of clusters according to their expression ratios, such that within each cluster, genes are coregulated or similarly expressed but have different expression levels when compared with genes of the other clusters. It is observed that each particular group of genes are expressed or not expressed subjected to the same environmental conditions or the same time-ranges. This *gene expression profiling* information is subsequently used for understanding functional pathways and how genes and gene products (say, proteins) interact with each other. This is referred to as the gene network analysis.

Microarray studies in the recent past have resulted in an *enormous* amount of gene expression data in the open literature, for several organisms under different experimental conditions of interest. The huge amount of gene expression data (as compared to traditional chemical engineering problems) makes it a challenging task to extract meaningful biological knowledge using mathematical and informatics tools. A seed-based NSGA-II was proposed and discussed by Garg and coworkers (Garg, 2009; Sikarwar, 2005) to group genes into various clusters based on microarray data. In their methodology, an MOO problem is defined with the goal of minimizing the intracluster distances while maximizing the intercluster distances. A small fraction of the GA population of size, N_p, is supplied as a seed population during initialization, based on empirical rules. This does not affect the diversity of the GA population, but provides "good" initial chromosomes (if the chromosomes are "bad" they will not be selected for the gene pool, probabilistically). This results in faster converge to the optimal solutions in the presence of noise in the big data-sets from microarray experiments. It is to be noted that the same seed-based adaptation of GA may be used fruitfully for solving other (much simpler) chemical engineering problems provided some empirical information for generating the seeds is available (see Ramteke and Gupta, 2009b for details on a seed-based adaptation based on the Haeckel-Baer biogenetic law of embryology).

For gene expression profiling, seed chromosomes are generated using a simple distance-based clustering on the gene expression data. The Euclidean distance between each gene is calculated as

$$d_{ab} = \sqrt{\sum_{k=1}^{m} (x_{ak} - x_{bk})^2}, \quad \forall a = 1, \ldots, (n-1) \text{ and } \quad b = (a+1), \ldots, n \quad a \neq b \qquad [4.11]$$

where x_{ij} is the gene expression ratio of the i-th gene in the k-th microarray experiment, m is the dimensionality of the experimental space (number of *distinct* experiments at which expression ratios are observed for each gene) and d_{ij} is the Euclidean distance between the i-th and j-th genes. These values are then mapped between 0 and 1 by using linear mapping

$$d_{ij} = \frac{(d_{ab} - d_{\min})}{(d_{\max} - d_{\min})}, \quad i \neq j, \quad \forall a,b \qquad [4.12]$$

where $i = 1, \ldots, (n-1), j = (i+1), \ldots, n$, and $d_{\min}$ and $d_{\max}$ are the overall minimum and maximum distances, respectively, between all genes being studied on the microarray. The normalized distance of each gene is compared with that for all the other genes. If the distances are less than a multiple of the average of $d_{\min}$ and $d_{\max}$, the genes are assigned to a single cluster. The process continues till all the genes are associated with at least one cluster. The average expression ratio of each cluster is then calculated on the basis of the association information. These calculated expression ratios are used as seed chromosomes in the GA population. A mixed population is generated for different values of the multiple of the average of $d_{\min}$ and $d_{\max}$, and used in GA. Results for a *simple* test case are illustrated in Fig. 4.7. Figure 4.7A shows the average target

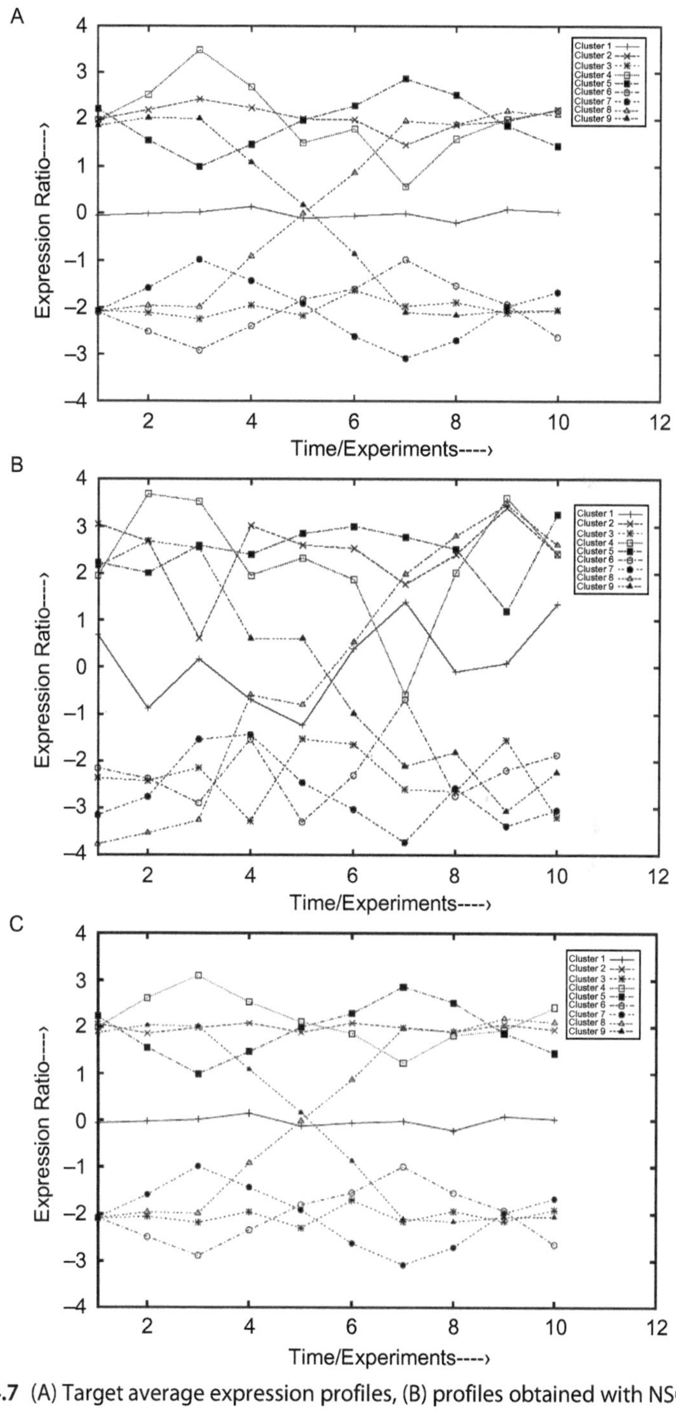

Figure 4.7 (A) Target average expression profiles, (B) profiles obtained with NSGA-II-JG, and (C) profiles obtained with seeded NSGA-II-JG. *Adapted from Garg (2009).*

profile of nine clusters across 10 different experiments based on their expressions, as observed from microarray experiments. Figure 4.7B shows the results of an MOO clustering, as discussed before, using NSGA-II-JG. It is clear that the code is not able to converge to the average expression profiles as shown in Fig. 4.7A. In contrast, the average expression profiles shown in Fig. 4.7C, using the seed-based NSGA-II-JG, match well with those shown in Fig. 4.7A. For more details and a *real-life* application, the readers are referred to Garg (2009) and Sikarwar (2005).

The Fortran 90 codes for some of the adaptations of NSGA-II (adapted from the *original* FORTRAN code of NSGA-II developed by Deb, http://www.iitk.ac.in/kangal/codes.shtml) are available on the following Website: http://www.iitk.ac.in/che/skg.html and at http://www.che.iitb.ac.in/faculty/skg.html. These codes can be modified for any future JG adaptations easily. The Websites of Deb as well as Gupta now have codes in C.

5. ALTRUISTIC ADAPTATION OF NSGA-II-aJG

While the several JG adaptations of NSGA-II have been in use over the last decade, other bio-mimetic adaptations of NSGA-II have also been proposed recently (and will continue to be developed in the future). One recent adaptation is the bio-mimetic adaptation based on altruism. The altruistic (Alt) behavior (Gadagkar, 1997) of honey bees has been the inspiration behind this recent adaptation of NSGA-II-aJG, namely, Alt-NSGA-II-aJG (Ramteke and Gupta, 2009c). Honey bees (and some other species like wasps, ants, etc.) are haplo-diploid in nature, unlike mammals which are diploid. That is, the males (drones) have n chromosomes while the females (queen and worker bees) have $2n$. The queen mother goes out once in her lifetime and mates with one (or more) drone. Let us assume for simplicity that she mates with a single drone. She can store and nourish the sperms, and can use them at will to produce daughter worker bees. Figure 4.8 shows how, of the $2n$ chromosomes in the daughters, n are identical to the n chromosomes of the father drone while the remaining n are randomly selected from among the $2n$ of the queen. The male off-springs have n chromosomes randomly selected from the $2n$ of the queen. The "inclusive" fitness (Gadagkar, 1997) of the honeycomb can be increased if the daughter worker bees bring up their sibling sisters (called altruistic behavior in evolutionary biology) rather than procreate and produce their own off-springs (selfish behavior). This phenomenon is used *as an inspiration* to develop the Alt-NSGA-II-aJG (Ramteke and Gupta, 2009c). A user-specified number

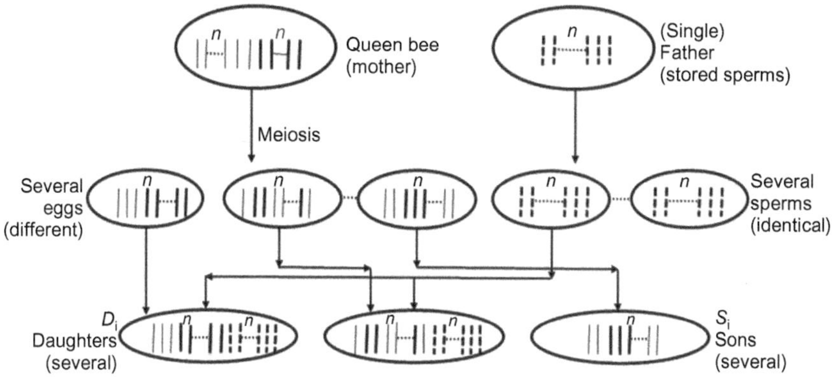

Figure 4.8 Chromosomes in the daughter (worker) and son (drone) bees.

of queens (better chromosomes, typically, a tenth of N_p) are used instead of a single one in the population of N_p solutions. In addition, three-point crossovers (Ramteke and Gupta, 2009c) instead of the two-point crossovers described earlier are used. This algorithm often, but not always, gives faster convergence than NSGA-II-aJG. It is clear that as for the JG adaptations, further improvements of the Alt-adaptation is required. Results are discussed later.

6. REAL-CODED GA (DEB, 2001)

A major difficulty in the binary-coded GA is what is referred to as Hamming's cliff: a transition from, say, string [0 1 1 1 1 1] to the next one, [1 0 0 0 0 0], involves the alteration of *several* binaries (by mutation). These lead to an artificial hindrance to a *gradual* search in the continuous search space. Real-coded GA has been suggested to overcome this and other problems. In this technique, real numbers are used to code the decision variables. Several crossover and mutation operators (Deb, 2001) are in use, and only one set is described here. Wright's (1991) crossover operator (between the j-th and k-th chromosomes selected randomly) for any decision variable, x_i, in any generation, generates three daughter chromosomes, $x_i^j - 0.5(x_i^k - x_i^j)$, $(x_i^j + x_i^k)/2$ and $x_i^k + 0.5(x_i^k - x_i^j)$, and selects the best two from these. Michalewicz's (1992) random mutation operator uses: $x_i^k + (R_i - 0.5)\Delta_i$, where Δ_i is the maximum perturbation permitted for x_i, and R_i is a random number in [0, 1].

Deb (2001) and coworkers reported the simulated binary crossover (SBX) operator that simulates the single point binary crossover operator in the real parameter space. Presently, this is one of the most commonly used real crossover operators in real-coded GAs. Moreover, they also reported a polynomial mutation operator for real-coded GAs using a polynomial function instead of a normal distribution function that is used in SBX. The readers are referred to Deb (2001) for more details.

7. BIO-MIMETIC RNA INTERFERENCE ADAPTATION

RNA interference (RNAi) is an evolutionary conserved mechanism of most eukaryotic cells that uses small, double-stranded RNA (dsRNA) molecules to direct sequence-dependent control of gene activity. In plants and lower organisms, RNAi also protects the genome from viruses and insertion of rogue genetic elements, for example, transposons (JG). The key event in the RNAi pathway is the generation of short dsRNA in the form of microRNA (miRNA) or small interfering RNA (siRNA). This short stranded RNA is a duplex with complementary strands, with only one of them participating in active silencing. RNA silencing has proved to be a useful technique with various applications including therapeutics, functional genomics, etc.

Here, we report a very recent (and still developing) use of an RNAi-based bio-mimetic adaptation of NSGA-II and NSGA-II-JG. This adaptation *preserves* the "good" chromosomes obtained using different GA operators, for example, crossover, mutation and JG, and prevents divergence over further generations, an unacceptable attribute of almost all GA codes. It is also noted that the use of the RNAi adaptation may result in a loss of heterogeneity in the population and may result in convergence to local solutions. This issue is addressed by using the RNAi adaptation on only a small fraction (8–10%) of the population (good chromosomes in terms of ranks, fitness values, etc.). The "proof of concept" is established using the ZDT4 test problem (Deb, 2001; discussed later). The results of the ZDT4 test problem are shown in Fig. 4.9. It is noted that the use of RNAi with elitism (Fig. 4.9C) provides a well spread and smooth Pareto front for the test problem at the 90th generation and, thus, may be more suitable for bio-informatics problems (like array informatics and gene network analyses). A short discussion of this very recent adaptation of NSGA has been included here just to indicate that attempts are continually being made to improve the speed

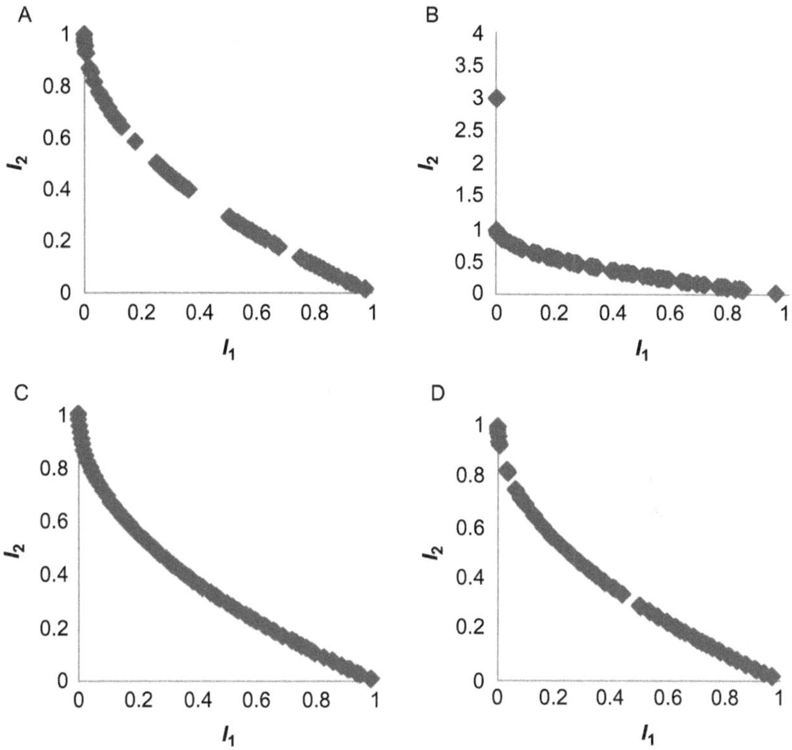

Figure 4.9 Comparison of the population at the 90th generation using (A) only elitism, no JG, no RNAi; (B) elitism and JG, no RNAi; (C) elitism and RNAi, no JG; and (D) Elitism, JG and RNAi.

of convergence of the GA codes so that better algorithms become available for compute-intense real-life problems.

8. SOME BENCHMARK PROBLEMS

The performance of the different algorithms can be tested using several benchmark problems, of which three are described below, namely, ZDT2, ZDT3, and ZDT4 (Deb, 2001). These are popular for testing new algorithms but are computationally far less intense than real-life Chemical Engineering MOO problems (see Jaimes and Coello Coello, 2009 who recommend researchers in other disciplines to use real-life applied problems for testing new algorithms rather than the simple benchmark problems).

Problem 1 (ZDT2)

$$\begin{aligned}
&\text{Min } I_1 = x_1 && \text{[a]}\\
&\text{Min } I_2 = g(\mathbf{x})\left\{1 - [x_1/g(\mathbf{x})]^2\right\} && \text{[b]}\\
&\text{where } g(\mathbf{x}) \text{ [the Rastrigin function] is given by} && \\
&g(\mathbf{x}) \equiv 1 + \frac{9}{n-1}\sum_{i=2}^{n} x_i && \text{[c]}\\
&\text{s.t.}: 0 \le x_j \le 1;\ j = 1, 2, \ldots, n && \text{[d]}
\end{aligned} \quad [4.13]$$

with $n = 30$. The Pareto-optimal front corresponds to $0 \le x_1 \le 1$, $x_j = 0$, $j = 2, 3, \ldots, 30$ ($0 \le I_1 \le 1$ and $0 \le I_2 \le 1$). The complexity of the problem lies in the fact that the Pareto front is nonconvex.

Problem 2 (ZDT3)

$$\begin{aligned}
&\text{Min } I_1 = x_1 && \text{[a]}\\
&\text{Min } I_2 = g(\mathbf{x})\left[1 - \{x_1/g(\mathbf{x})\}^{1/2} - \{x_1/g(\mathbf{x})\}\sin(10\pi x_1)\right] && \text{[b]}\\
&g(\mathbf{x}) \equiv 1 + \frac{9}{n-1}\sum_{i=2}^{n} x_i && \text{[c]}\\
&\text{s.t.}: 0 \le x_i \le 1;\quad i = 1, 2, \ldots, n && \text{[d]}
\end{aligned}$$

$$[4.14]$$

with $n = 30$. The Pareto-optimal front corresponds to $x_j = 0, j = 2, 3, \ldots, 30$. This problem is a good test for any MOO algorithm since the Pareto front is discontinuous.

Problem 3 (ZDT4)

$$\begin{aligned}
&\text{Min } I_1 = x_1 && \text{[a]}\\
&\text{Min } I_2 = g(\mathbf{x})\left[1 - [x_1/g(\mathbf{x})]^{1/2}\right] && \text{[b]}\\
&g(\mathbf{x}) \equiv 1 + 10(n-1) + \sum_{i=2}^{n}\{x_i^2 - 10\cos(4\pi x_i)\} && \text{[c]}\\
&\text{s.t.}: 0 \le x_1 \le 1 && \text{[d]}\\
&-5 \le x_j \le 5;\quad j = 2, 3, \ldots, n && \text{[e]}
\end{aligned} \quad [4.15]$$

with $n = 10$. This problem has 99 Pareto fronts, of which only one is the *global* optimal. The latter corresponds to $0 \le x_1 \le 1$, $x_j = 0, j = 2, 3, \ldots, 10$ and so $0 \le I_1 \le 1$ and $0 \le I_2 \le 1$.

The binary-coded NSGA-II as well as PAES (Knowles and Corne, 2000) have been found (Deb, 2001) to converge to *local* Paretos, rather than to the

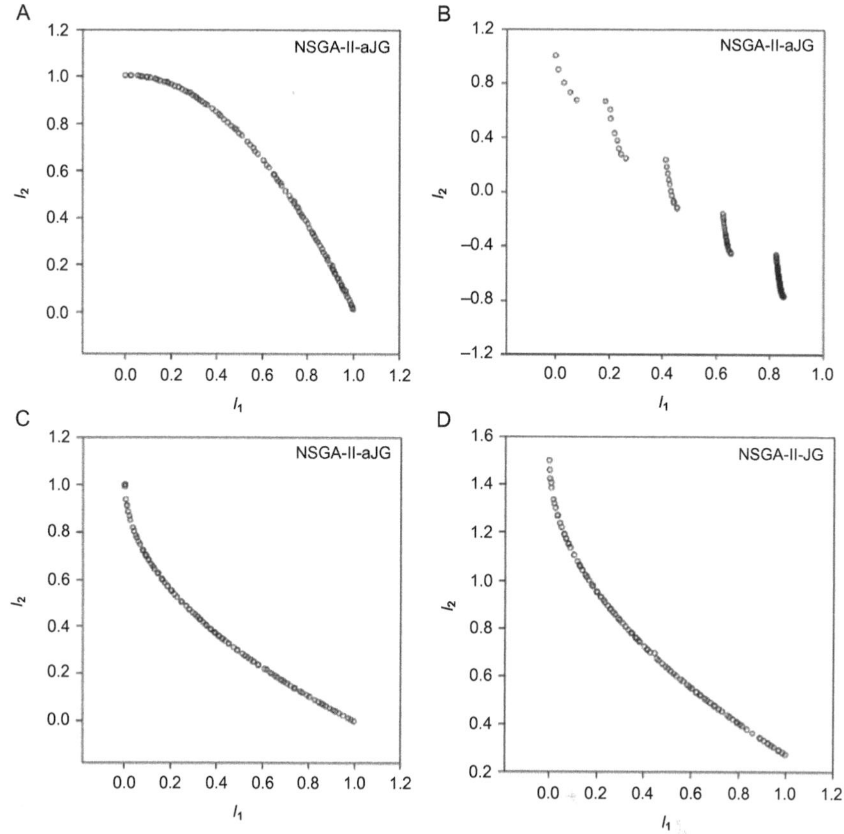

Figure 4.10 Optimal solutions for (A) Problem 1 (ZDT2, Eq. 4.13), (B) Problem 2 (ZDT3, Eq. 4.14), (C) Problem 3 (ZDT4, Eq. 4.15) using NSGA-II-aJG, and for (D) Problem 3 (ZDT4, Eq. 4.15) using NSGA-II. $N_{gen}=1000$. *Adapted from Agarwal (2007).*

global optimal set (the *real-coded* NSGA-II, discussed earlier, has been found to converge to the global Pareto set, though in 100,000 function evaluations).

The three benchmark problems are solved using NSGA-II-aJG. The *best* values of the computational parameters are found by trial (this is a big irritant in GA, particularly for compute-intense real-life problems) for the three problems. These are given in Table 4.1. Figure 4.10A–C (Agarwal, 2007) give the results using this JG adaptation at the end of 1000 generations. Figure 4.10D shows the solutions using NSGA-II-JG at the end of 1000 generations (involving the same computational effort) for Problem 3. It is observed that we obtain a local Pareto set with the latter technique (note the value of I_2 is above the correct maximum value of 1.0). It may be mentioned that the binary-coded NSGA-II-JG *does* give the correct Pareto solution for Problem 3 but only at about $N_{gen}=1600$ (but the binary-coded NSGA-II does not converge at all for this problem even after 400,000 function evaluations). Correct Pareto sets are also obtained using NSGA-II-sJG and NSGA-II-saJG (Agarwal and Gupta, 2008a) for all three problems with $N_{\text{gen}} = 1000$, as well as by using NSGA-II-JG for Problems 1 and 2 (but not for Problem 3).

9. SOME METRICS FOR COMPARING PARETO SOLUTIONS

Pareto-optimal solutions need to be compared, particularly for real-life MOO problems, using *quantitative*, nonvisual methods (called metrics). Several metrics (Deb, 2001) have been proposed and some are described here.

a. The set-coverage matrix, **C**: The elements, $C_{p,q}$, of the set-coverage matrix for techniques p and q represent the fraction of solutions obtained using technique, q, that are (weakly) dominated by the solutions obtained using technique, p. For example, $C_{2,1}$ (when 2 is NSGA-II-aJG and 1 is NSGA-II-JG) for the ZDT4 problem (see Table 4.2) is 0.99. This means that almost all solutions obtained using NSGA-II-aJG are better than those obtained with NSGA-II.

b. The maximum spread, MS: The maximum spread is the length of the diagonal of the hyper-box formed by the extreme function values in the nondominated set. For two-objective problems, this metric refers to the Euclidean distance between the two extreme solutions in the **I** space. For the ZDT4 problem, Table 4.2 shows that the MS is higher for NSGA-II-JG than for NSGA-II-aJG and so the former is better.

c. The "spacing," S: The spacing is a measure of the relative distance between *consecutive* (nearest neighbor) solutions in the nondominated set (really speaking, it is the standard deviation of the nearest neighboring distances). It is given by

$$S = \sqrt{\frac{1}{Q}\sum_{i=1}^{Q}(d_i - \bar{d})^2} \quad [a]$$

where

$$d_i = \min_{k \in Q; k \neq i} \sum_{l=1}^{m} \left| I_l^i - I_l^k \right| \quad [b] \qquad [4.16]$$

$$\bar{d} = \sum_{i=1}^{Q} \frac{d_i}{Q} \quad [c]$$

Table 4.2 Metrics for Problems 1–3 (Agarwal and Gupta, 2007) with NSGA-II-JG and NSGA-II-aJG after 1000 generations

	NSGA-II-JG	NSGA-II-aJG
Problem 1 (ZDT2)		
Set coverage metric		
NSGA-II-JG	–	1.60×10^{-1}
NSGA-II-aJG	2.20×10^{-1}	–
Spacing	8.66×10^{-3}	7.18×10^{-3}
Maximum spread	1.4004	1.4091
Problem 2 (ZDT3)		
Set coverage metric		
NSGA-II-JG	–	1.00×10^{-2}
NSGA-II-aJG	4.30×10^{-1}	–
Spacing	2.25×10^{-2}	2.55×10^{-2}
Maximum spread	1.9722	1.9692
Problem 3 (ZDT4)		
Set coverage metric		
NSGA-II-JG	–	0
NSGA-II-aJG	9.90×10^{-1}	–
Spacing	9.27×10^{-3}	7.74×10^{-3}
Maximum spread	1.5809	1.4138

In Eq. (4.16), m is the number of objective functions and Q is the number of nondominated points. Clearly, d_i is the "spacing" (sum of *each* of the distances in the **I** space) between the i-th point and its *nearest* neighbors, $\bar{d}$ is its mean value, and S is the standard deviation of the different d_i. An algorithm that gives nondominated solutions having a smaller value of S but larger values of MS is, obviously, superior.

The three metrics discussed above are given for Problems 1–3 in Table 4.2 for two techniques, NSGA-II-JG and NSGA-II-aJG. It is observed for the ZDT4 problem that NSGA-II-JG gives a high value of MS (desirable) but performs poorly in terms of set-coverage and spacing, and so is inferior. Again, which adaptation performs better is problem-specific.

d. Box plots (Chambers et al., 1983): Yet another method to compare algorithms for MOO problems is the box plots (Chambers et al., 1983). These are shown for Problems 1–3 in Fig. 4.11, not only for NSGA-II-JG and NSGA-II-aJG but for NSGA-II-saJG and NSGA-II-sJG as well. These plots show the distribution (in terms of quartiles and outliers) of the points, graphically. For example, the box plot of I_1 for any technique indicates the entire range of I_1 distributed over four quartiles, with 0–25% of the solutions having the lowest values of I_1 indicated by the lower vertical line with a whisker (except for outliers, see later), the next 25–50% of the solutions by the lower box, 50–75% of the solutions by the upper part of the box, and the remaining 75–100% of the solutions having the highest values (except for outliers) of I_1, by the upper vertical line with a whisker. Points beyond the 5% and 95% range (outliers) are shown by separate circles. The mean values of **I** are shown by dotted lines inside the boxes. A good algorithm should give box plots in which all the regions are equally long, and the mean line coincides with the upper line of the lower box. It is observed that for Problem 1, NSGA-II-sJG gives the best box plot. For Problem 2, NSGA-II-aJG gives the best box plot; while for Problem 3, NSGA-II-sJG and NSGA-II-saJG give comparable results. Clearly, the performance of the algorithms is problem-specific. A study of all the results indicates that NSGA-II-JG is inferior to the other algorithms, at least for the three benchmark problems studied. NSGA-II-sJG and NSGA-II-saJG appear to be satisfactory and comparable. The latter two algorithms do not have the disadvantage of user-defined fixed length of the JG, as required in NSGA-II-aJG.

e. One may get an idea of the value of N_{gen} at which computations may be terminated (of course, this needs obtaining the converged "optimal" results at high values of N_{gen}) by evaluating

$$\sigma^2 = \frac{\sum_{j=2}^{N}\sum_{i=1}^{N_p}\left(\frac{I_{j,i} - I_{j,\text{opt},i}}{\text{Range of } I_{j,\text{opt}}}\right)^2}{(N-1)N_p} \qquad [4.17]$$

for each generation. To evaluate σ^2 for an N-objective MOO problem, we select, say the i-th point, $I_{1,\text{opt},i}$, $I_{2,\text{opt},i}$, . . . , $I_{N,\text{opt},i}$, on the converged (final) Pareto set of N_p points. Thus, $I_{j,\text{opt},i}$ is the value of the j-th objective function, I_j, for the i-th point in the final Pareto solution. $I_{j,i}$ is the *interpolated*

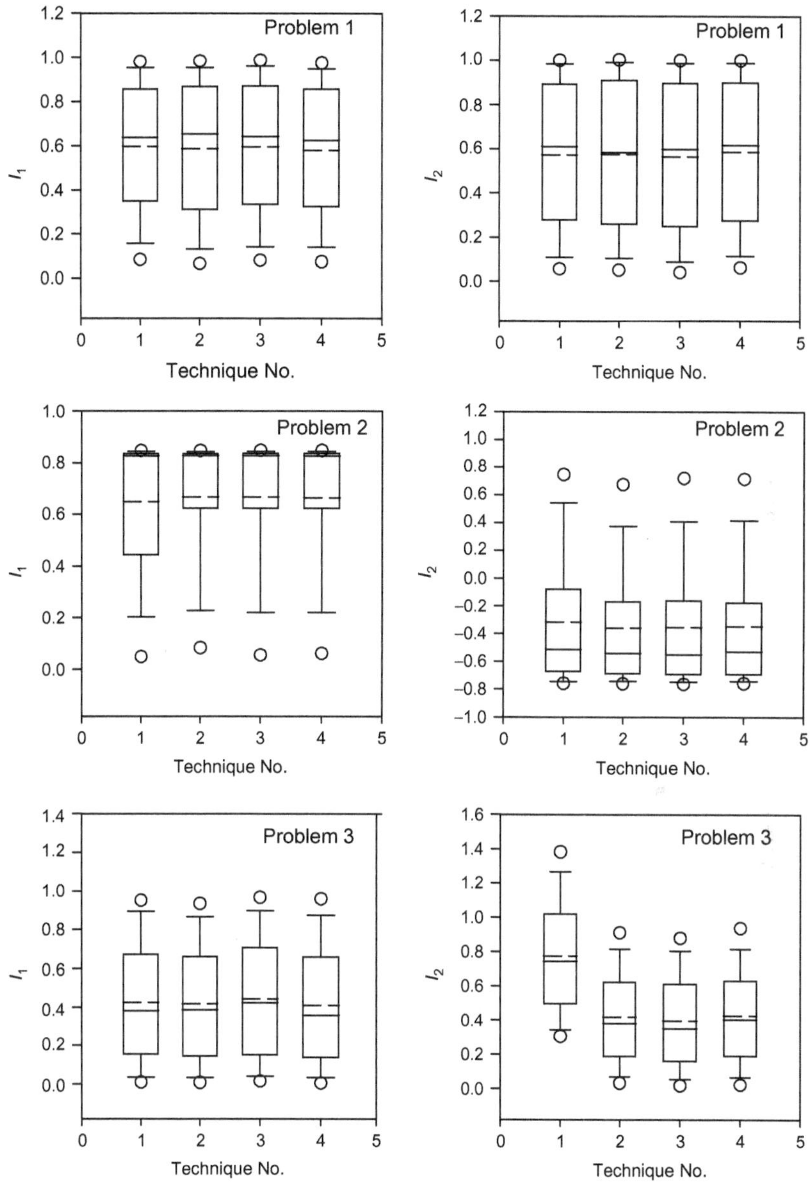

Figure 4.11 Box plots of I_1 and I_2 for Problems 1 (ZDT2), 2 (ZDT3) and 3 (ZDT4) after 1000 generations. Technique 1: NSGA-II-JG, technique 2: NSGA-II-aJG, technique 3: NSGA-II-saJG, and technique 4: NSGA-II-sJG. *Adapted from Agarwal (2007).*

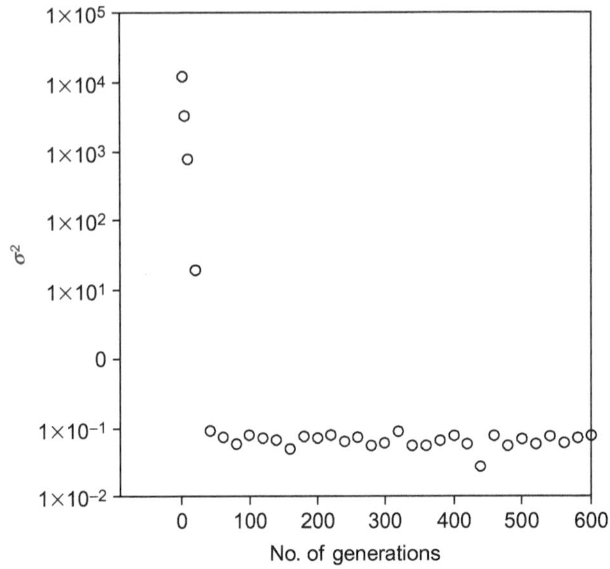

Figure 4.12 Results for Alt-NSGA-II-aJG for the ZDT4 problem. *Adapted from Ramteke and Gupta (2009c).*

value of I_j at an *earlier* (nonconverged) generation corresponding to point $I_{1,\mathrm{opt},i}$. σ^2 will be unity for the converged Pareto set and will be higher at the earlier generations. Values of σ^2 below about 0.1 indicate convergence. Figure 4.12 shows (Ramteke and Gupta, 2009b) the rapid decrease of σ^2 to below 0.1 for the ZDT4 problem, indicating the superiority of Alt-NSGA-II-aJG.

10. SOME CHEMICAL ENGINEERING APPLICATIONS

Some real-life MOO examples from the domain of Chemical Engineering are now described. These are better tests of MOO algorithms than the benchmark problems discussed earlier.

10.1. MOO of heat exchanger networks

Agarwal and Gupta (2008b) developed and used NSGA-II-sJG/saJG for optimizing HENs. Figure 4.13 shows a typical HEN with three hot process streams (upper three horizontal lines with arrows pointed to the right) and three cold streams (lower three horizontal lines with the arrows pointed towards the left). Complete details are given in Agarwal and Gupta (2008b). Intermediate heat exchange between the hot and cold process streams (and using additional hot

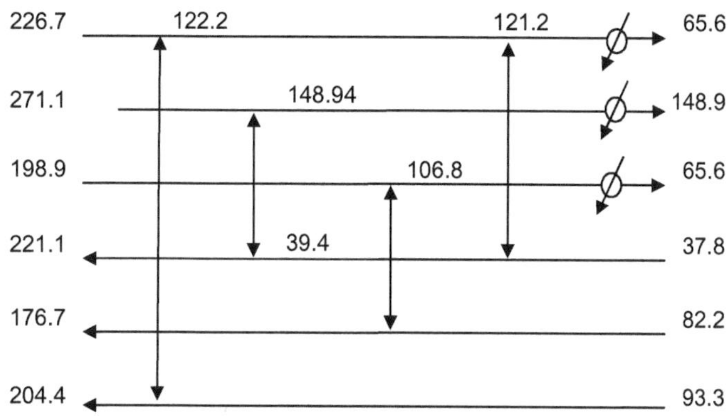

Figure 4.13 Three hot and three cold process streams with optimal values of the intermediate temperatures (and utilities) indicated. *Adapted from Agarwal (2007).*

and cold utilities, if required) need to be done so as to maximize the energy efficiency of the network. The optimal *number* of intermediate heat exchanges and the optimal values of the intermediate temperatures are the decision variables for this problem. The length of the chromosomes in the population can differ since the number of intermediate temperatures in any stream can be different. This requires an adaptation of the GA algorithm and is described in Agarwal and Gupta (2008b).

Several interesting MOO problems for HENs have been solved. One is given below

$$\begin{aligned} &\text{Min } I_1 = \text{Annual cost} && [a] \\ &\text{Min } I_2 = \text{Total hot and cold utility requirements (kW)} && [b] \qquad [4.18] \\ &\text{s.t. constraints and bounds} && [c] \end{aligned}$$

Reducing the total requirement of hot and cold utilities is important for the conservation of water, a natural resource. The single-objective results (minimizing the total cost of the HEN) for this system using the *heuristic* approach of Linnhoff and Ahmad (1990) are shown by a filled square in Fig. 4.14. This diagram also shows the results of the MOO problem (Eq. 4.18) for this system. It is observed that one can reduce the total utility requirement from about 58,000 kW for the single-objective solution (min cost) to about 50,000 kW with only a small increase in the cost. The usefulness of MOO and the concept of trade-off is quite well illustrated in Fig. 4.14. It may be mentioned that the optimal number of intermediate temperatures (HXs) in each stream are not specified *a priori*. The first few substrings of a chromosome are used for the values (integral) of the HXs in each stream. This is one of the problem-specific "tricks" mentioned earlier.

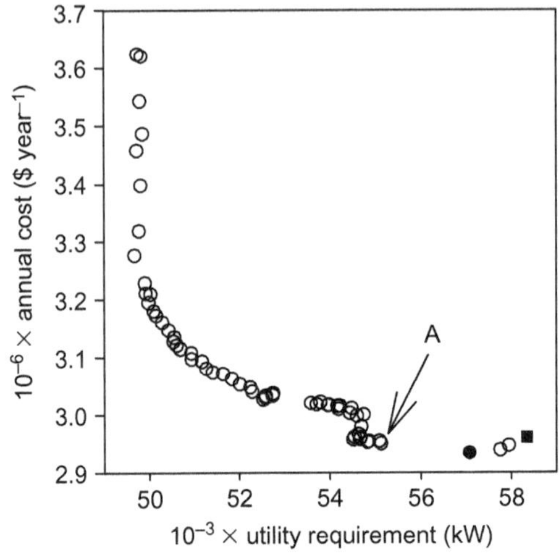

Figure 4.14 Optimal Pareto front for Eq. (4.18). ■, SOO solution of Linnhoff and Ahmad (1990); ●, SOO results of Agarwal and Gupta (2007, 2008b). *Adapted from Agarwal (2007).*

10.2. MOO of a catalytic fixed-bed maleic anhydride reactor

The kinetic scheme of the highly exothermic catalytic production of maleic anhydride (MA) is shown below (Chaudhari and Gupta, 2012)

$$C_4H_{10}(Bu) + 3.5O_2 \xrightarrow{k_1} C_4H_2O_3(MA) + 4H_2O \text{ (desired reaction)} \qquad [a]$$
$$C_4H_2O_3 + pO_2 \xrightarrow{k_2} (6-2p)CO + (2p-2)CO_2 + H_2O \text{ (decomposition)} \qquad [b]$$
$$C_4H_{10} + nO_2 \xrightarrow{k_3} (13-2n)CO + (2n-9)CO_2 + 5H_2O \text{ (total oxidation)} \qquad [c]$$
$$[4.19]$$

A fixed-bed catalytic reactor has been modeled incorporating the diffusion of this seven-component system into the pores of the cylindrical catalyst particles. Some of the model-parameters are tuned using a set of data on pilot plants as well as on an industrial reactor. The model is then used to solve MOO problems. One of the three-objective optimization problems that have been solved using NSGA-II-aJG is

$$\begin{aligned}
&\text{Max}: F_1\left(G^0, y^0_{Bu}, P^0_T, T^0, T_S\right) \equiv F_{MA}(\text{kmol/s}) && [a]\\
&\text{Min}: I_2\left(G^0, y^0_{Bu}, P^0_T, T^0, T_S\right) \equiv F^0_{Bu}(\text{kmol/s}) && [b]\\
&\text{Min}: I_3\left(G^0, y^0_{Bu}, P^0_T, T^0, T_S\right) \equiv F_{CO} + F_{CO_2}(\text{kmol/s}) && [c]\\
&\text{s.t. bounds and constraints (Chaudhari and Gupta, 2012)} && [d]
\end{aligned} \qquad [4.20]$$

In Eq. (4.20), F_{MA} is the exit flow rate of the (desired) MA, F^0_{Bu} is the feed flow rate of *n*-butane, while $F_{CO} + F_{CO_2}$ is the flow rate of the undesirable carbon oxides. The decision variables are G^0, superficial mass velocity of gas at the inlet; y^0_{Bu}, mole fraction of *n*-butane in the inlet stream; P^0_T, total pressure at the inlet; T^0, temperature of the inlet stream; and T_S, coolant temperature. The set of $N_p = 60$ nondominated solutions is shown in Fig. 4.15. Figure 4.15A shows the solutions in terms of reordered chromosome numbers so that F_1 is arranged in increasing order. Figure 4.15B and C show the other two-objective functions using the same (new) chromosome numbers as in Fig. 4.15A. This method of plotting the optimal solutions is easier to interpret and can be used for problems involving more than two or three objectives. It is clear that F_1 improves, but I_2 and I_3 both worsen simultaneously, indicating a Pareto-kind behavior. It is also found (results not shown) that the altruistic adaptation, Alt-NSGA-II-aJG, converges to the optimal solutions faster for two-objective optimization problems, but is slower than NSGA-II-aJG for three-objective problems. A further adaptation of NSGA-II-aJG was developed for the three-objective optimization problem (Eq. 4.20) to replace optimal points associated with extreme sensitivity and simultaneously give smoother Pareto sets (this is one of several problem-specific "tricks" referred to earlier).

10.3. Summary of some other MOO problems

Batch reactor studies are common in Chemical Engineering and are associated with the use of continuous decision variables. The MOO of these problems involves obtaining optimal values of the *histories* of the decision variables (as optimal *functions* of time instead of optimal *values* as for problems discussed earlier). These are referred to as *trajectory* optimization problems (akin to optimal trajectories required in aerospace engineering). Mitra et al. (1998) developed the procedure to solve such problems (again a problem-specific "trick") using NSGA-I, an earlier version of the elitist NSGA-II. They optimized an industrial nylon-6 reactor (see Fig. 4.16) modeled by Wajge et al. (1994). In their study, one decision variable [the rate of release, $V_T(t)$, of vapor from the reactor through the control valve] was a continuous *function* of the reaction time, *t*, while the other, the temperature, T_J, of the jacket fluid, was an optimal *value* (a number). More recently, Ramteke and Gupta (2008) carried out the MOO of the same industrial nylon-6 reactor using both $V_T(t)$ and $T_J(t)$ as decision variables. They obtained the two optimal trajectories using NSGA-II-aJG. The use of trajectories of two decision variables leads to better solutions as compared to results with a single trajectory, $V_T(t)$, only.

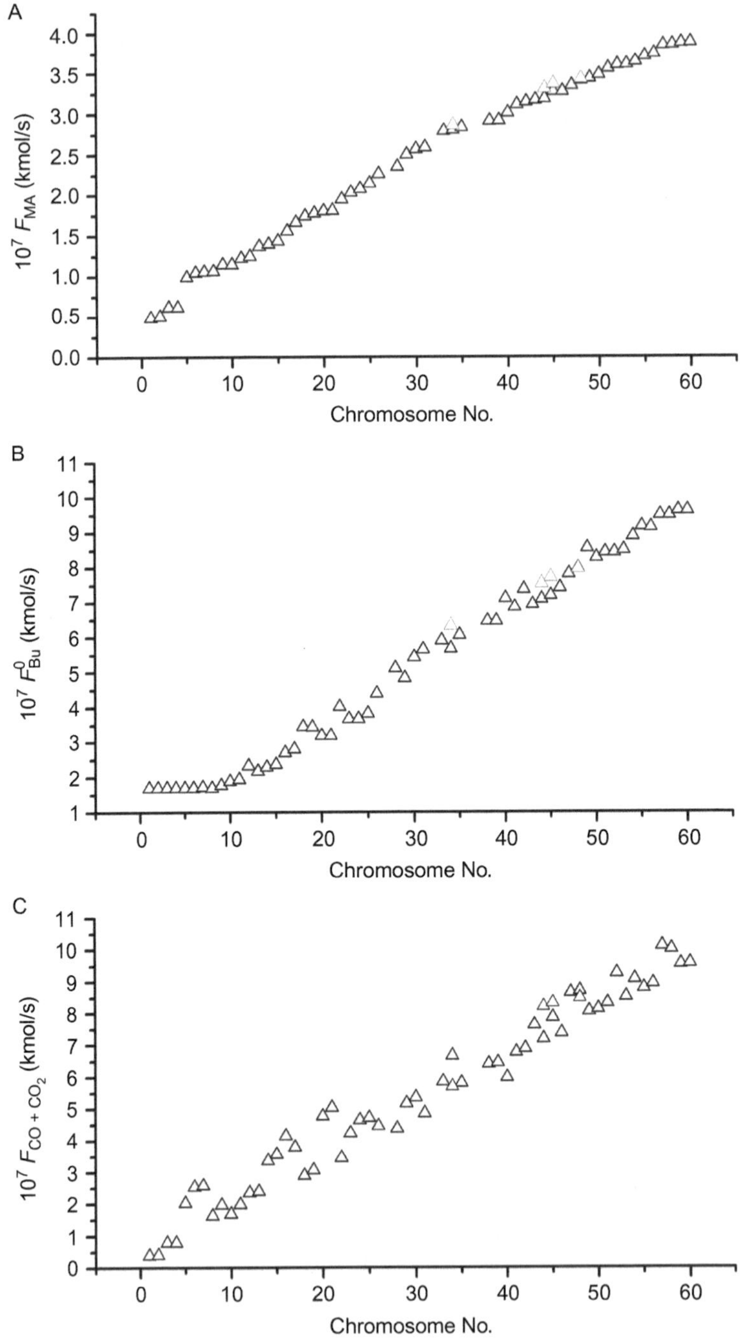

Figure 4.15 Three-objective optimization results of Eq. (4.20) for maleic anhydride (for one of the pilot plant reactor systems). (A) I_1 (in increasing order), (B) corresponding values of I_2, and (C) I_3. *Adapted from Chaudhari and Gupta (2012).*

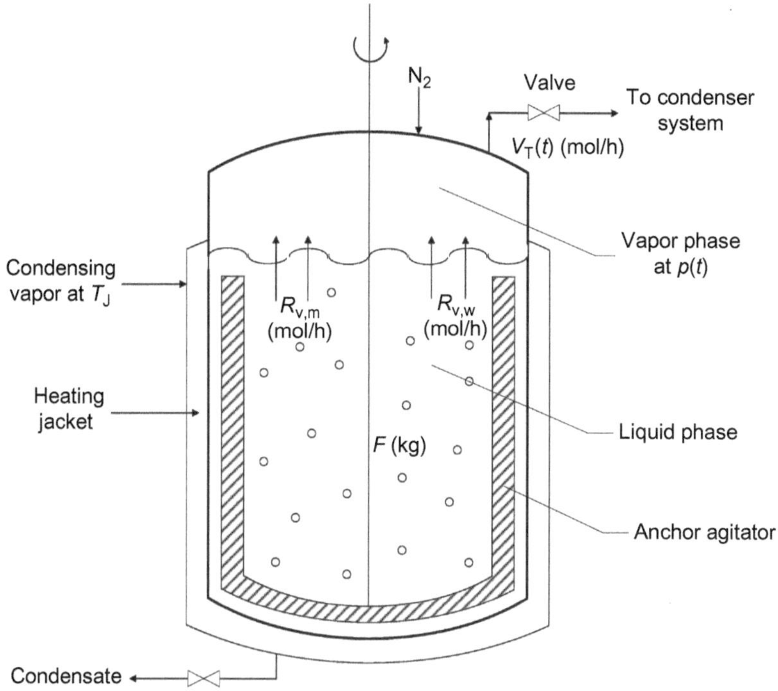

Figure 4.16 Schematic of an industrial nylon-6 semibatch reactor. *Adapted from Ramteke and Gupta (2008).*

Yuen et al. (2000) carried out the MOO of a membrane separation unit for the production of low alcohol beer having a good "taste." They used NSGA-I. Guria et al. (2005a) used NSGA-II-aJG for the MOO of membrane-based water desalination units. Guria et al. (2005b) later developed and used NSGA-II-mJG for the optimization of froth flotation circuits. Industrial steam reformers, both under steady operation (Rajesh et al., 2000) and under unsteady conditions (Nandasana et al., 2003) to counter the effect of disturbances, were optimized using multiple objectives [Rajesh et al., 2000 developed a procedure ("trick") for making the bounds of *some* of the decision variables dependent on the mapped values of some of the other decision variables]. Similarly, MOO of an industrial FCCU (Kasat et al., 2002; see Fig. 4.2 for a Pareto-optimal solution), and of a pressure swing adsorption unit (Sankararao and Gupta, 2007b) have also been carried out. A nine catalyst-zone phthalic anhydride reactor (see Fig. 4.17) has been multiobjectively optimized by Bhat and Gupta (2008) and by Ramteke and Gupta (2009c). The latter found that Alt-NSGA-II-aJG performed better (see Fig. 4.18A) than NSGA-II-aJG. Bhat et al. (2006) and Bhat (2007) used NSGA-II-aJG for the *experimental* on-line optimizing control

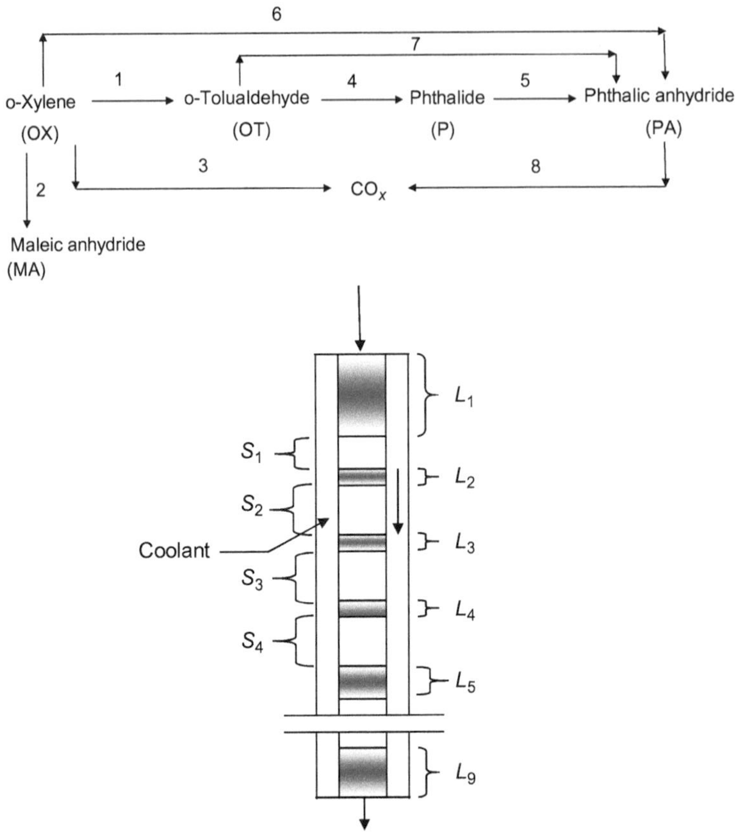

Figure 4.17 Kinetic scheme for phthalic anhydride manufacture and a schematic of the present-day nine-zone reactor. *Adapted from Ramteke and Gupta (2009c).*

of poly-methyl methacrylate polymerization in a 1-l batch reactor, with a simulated disturbance (electrical power switched off for some time). Bhaskar et al. (2000b) carried out the MOO of an industrial wiped-film poly ethylene terephthalate (PET) reactor. These workers got single optimal solutions and had to generate several of these solutions using different values of the computational parameters to generate the entire Pareto set (a "trick"). Agarwal et al. (2007) reported a similar study for an industrial low density polyethylene (LDPE) reactor with multiple (intermediate) injections of the initiator. Kundu et al. (2009), Wongsu et al. (2004), Yu et al. (2003), and Zhang et al. (2002a,b, 2003a,b, 2004) have carried out the MOO of simulated moving bed (SMB) chromatographic reactor. The reader is referred to the original papers for details. We would like to mention that each of these studies require small modifications (we call them "tricks")

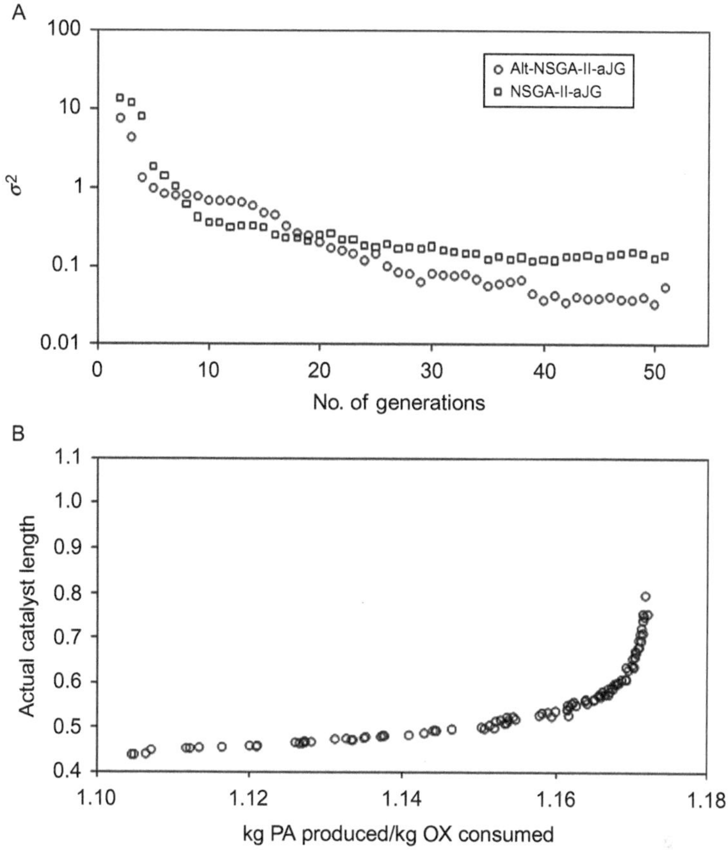

Figure 4.18 Optimal solutions for the nine-zone phthalic anhydride (PA) reactor. Max $F_1 \equiv$ kg PA produced/kg o-xylene consumed; Min $I_2 \equiv$ total length of (actual) catalyst. *Adapted from Ramteke and Gupta (2009c).*

of NSGA-II and its several bio-mimetic adaptations to get optimal Pareto solutions and exposure to several of these would enable the readers to develop their own "tricks."

11. CONCLUSIONS

MO GA is an extremely popular evolutionary optimization technique for solving problems involving two or more objective functions. Such MO optimizations are far more meaningful and relevant for industrial problems, and are important in these days of intense competition. Usually, one obtains sets of several equally good (nondominated) Pareto-optimal solutions. One of the MOGA algorithms is the elitist NSGA-II. Unfortunately, most MOGA codes, including NSGA-II, are extremely slow when applied to *real-life* problems. Several bio-mimetic adaptations have been described which improve the rates of convergence. A few chemical engineering examples involving two or three noncommensurate objective functions are described. These include HENs, industrial catalytic reactors for the manufacture of maleic anhydride and phthalic anhydride, industrial third stage polyester reactors, LDPE reactors with multiple injections of initiator, an industrial semibatch nylon-6 reactor, etc. A seed-based adaptation is discussed which helps obtain faster converged solutions for highly compute-intense problems in bioinformatics (e.g., clustering of data from cDNA microarray experiments). A *very* recent RNAi adaptation of NSGA-II is presented which holds promise for greatly improved rates of convergence.

It may be added in the end that the future developments in this area would take place along two lines: (a) development of newer adaptations and "tricks" which would lead to faster convergence and (b) the solution of still more compute-intense real-life problems as the computational resources increase.

REFERENCES

Agarwal A: Multi-objective optimal design of heat exchangers and heat exchanger networks using new adaptations of NSGA-II. M.Tech. Thesis, Indian Institute of Technology, Kanpur, 2007.

Agarwal A, Gupta SK: Jumping gene adaptations of NSGA-II and their use in the multi-objective optimal design of shell and tube heat exchangers, *Chem Eng Res Des* 86:123–139, 2008a.

Agarwal A, Gupta SK: Multiobjective optimal design of heat exchanger networks using new adaptations of the elitist nondominated sorting genetic algorithm, NSGA-II, *Indus Eng Chem Res* 47:3489–3501, 2008b.

Agarwal N, Rangaiah GP, Ray AK, Gupta SK: Design stage optimization of an industrial low-density polyethylene tubular reactor for multiple objectives using NSGA-II and its jumping gene adaptations, *Chem Eng Sci* 62:2346–2365, 2007.

Beveridge GSG, Schechter RS: *Optimization: theory and practice,* New York, 1970, McGraw Hill.

Bhaskar V, Gupta SK, Ray AK: Applications of multiobjective optimization in chemical engineering, *Rev Chem Eng* 16:1–54, 2000a.

Bhaskar V, Gupta SK, Ray AK: Multiobjective optimization of an industrial wiped film poly (ethylene terephthalate) reactor, *AIChE J* 46:1046–1058, 2000b.

Bhat GR, Gupta SK: MO optimization of phthalic anhydride industrial catalytic reactors using *guided* GA with the adapted jumping gene operator, *Chem Eng Res Des* 86:959–976, 2008.

Bhat SA, Gupta S, Saraf DN, Gupta SK: On-line optimizing control of bulk free radical polymerization reactors under temporary loss of temperature regulation: an experimental study on a 1-liter batch reactor, *Indus Eng Chem Res* 45:7530–7539, 2006.

Bhat SA: On-line optimizing control of bulk free radical polymerization of methyl methacrylate in a batch reactor using virtual instrumentation. Ph.D. Thesis, Indian Institute of Technology, Kanpur, 2007.

Bryson AE, Ho YC: *Applied optimal control,* Waltham, MA, 1969, Blaisdell.

Chambers JM, Cleveland WS, Kleiner B, Tukey PA: *Graphical methods for data analysis,* Belmont, CA, 1983, Wadsworth.

Chan TM, Man KF, Tang KS, Kwong S: A jumping gene algorithm for multiobjective resource management in wideband CDMA systems, *Comput J* 48:749–768, 2005a.

Chan TM, Man KF, Tang KS, Kwong S: Optimization of wireless local area network in IC factory using a jumping-gene paradigm. In *3rd IEEE international conference on industrial informatics (INDIN)*, 2005b, pp 773–778.

Chaudhari P, Gupta SK: Multi-objective optimization of a fixed bed maleic anhydride reactor using an improved biomimetic adaptation of NSGA-II, *Indus Eng Chem Res* 51:3279–3294, 2012.

Coello Coello CA, Veldhuizen DAV, Lamont GB: *Evolutionary algorithms for solving multi-objective problems,* ed 2, New York, 2007, Springer.

Deb K: *Optimization for engineering design: algorithms and examples,* New Delhi, India, 1995, Prentice Hall of India.

Deb K: *Multi-objective optimization using evolutionary algorithms,* Chichester, UK, 2001, Wiley.

Deb K, Pratap A, Agarwal S, Meyarivan TA: Fast and elitist multi-objective genetic algorithm: NSGA-II, *IEEE Trans Evol Comput* 6:181–197, 2002.

Edgar TF, Himmelblau DM, Lasdon LS: *Optimization of chemical processes,* ed 2, New York, 2001, McGraw Hill.

Gadagkar R: *Survival strategies of animals: cooperation and conflicts,* Cambridge, MA, 1997, Harvard University Press.

Garg S: Array informatics using multi-objective genetic algorithms: from gene expressions to gene networks. In Rangaiah GP, editor: *Multi-objective optimization: techniques and applications in chemical engineering*, Singapore, 2009, World Scientific, pp 363–400.

Gill PE, Murray W, Wright MH: *Practical optimization,* New York, 1981, Academic.

Goldberg DE: *Genetic algorithms in search, optimization and machine learning,* Reading, MA, 1989, Addison-Wesley.

Guria C, Bhattacharya PK, Gupta SK: Multi-objective optimization of reverse osmosis desalination units using different adaptations of the non-dominated sorting genetic algorithm (NSGA), *Comp Chem Eng* 29:1977–1995, 2005a.

Guria C, Verma M, Mehrotra SP, Gupta SK: Multi-objective optimal synthesis and design of froth flotation circuits for mineral processing using the jumping gene adaptation of genetic algorithm, *Indus Eng Chem Res* 44:2621–2633, 2005b.

Haimes YY: *Hierarchical analysis of water resources systems,* New York, 1977, McGraw Hill.

Haimes YY, Hall WA: Multiobjectives in water resources systems analaysis: the surrogate worth trade-off method, *Water Resources Res* 10:615–624, 1974.

Holland JH: *Adaptation in natural and artificial systems,* Ann Arbor, MI, 1975, University of Michigan Press.

Jaimes AL, Coello Coello CA: Multi-objective evolutionary algorithms: a review of the state-of-the-art and some of their applications in chemical engineering. In Rangaiah GP, editor: *Multi-objective optimization: techniques and applications in chemical engineering*, Singapore, 2009, World Scientific, pp 61–90.

Kasat RB, Gupta SK: Multi-objective optimization of an industrial fluidized-bed catalytic cracking unit (FCCU) using genetic algorithm (GA) with the jumping genes operator, *Comput Chem Eng* 27:1785–1800, 2003.

Kasat RB, Kunzru D, Saraf DN, Gupta SK: Multiobjective optimization of industrial FCC units using *elitist* non-dominated sorting genetic algorithm, *Indus Eng Chem Res* 41:4765–4776, 2002.

Khosla DK, Gupta SK, Saraf DN: Multi-objective optimization of fuel oil blending using the jumping gene adaptation of genetic algorithm, *Fuel Proc Technol* 88:51–63, 2007.
Knowles JD, Corne DW: Approximating the non-dominated front using the Pareto archived evolution strategy, *Evol Comput* 8:149–172, 2000.
Kundu P, Zhang Y, Ray AK: Multiobjective optimization of oxidative coupling of methane in a simulated moving reactor, *Chem Eng Sci* 64:4137–4149, 2009.
Lapidus L, Luus R: *Optimal control of engineering processes,* Waltham, MA, 1967, Blaisdell.
Linnhoff B, Ahmad S: Cost optimum heat exchanger networks—1. Minimum energy and capital using simple models for capital cost, *Comp Chem Eng* 14:729–750, 1990.
Man KF, Chan TM, Tang KS, Kwong S: Jumping genes in evolutionary computing. In *The 30th annual conference of IEEE industrial electronics society (IECON'04), Busan, Korea*, 2004.
McClintock B: *The discovery and characterization of transposable elements: the collected papers of Barbara McClintock,* New York, 1987, Garland.
Michalewicz Z: *Genetic algorithms + data structures = evolution programs,* Berlin, 1992, Springer.
Mitra K, Deb K, Gupta SK: Multiobjective dynamic optimization of an industrial nylon 6 semibatch reactor using genetic algorithm, *J Appl Polym Sci* 69:69–87, 1998.
Nandasana AD, Ray AK, Gupta SK: Dynamic model of an industrial steam reformer and its use for multiobjective optimization, *Indus Eng Chem Res* 42:4028–4042, 2003.
Rajesh JK, Gupta SK, Rangaiah GP, Ray AK: Multiobjective optimization of steam reformer performance using genetic algorithm, *Indus Eng Chem Res* 39:706–717, 2000.
Ramteke M, Gupta SK: Multi-objective optimization of an industrial nylon-6 semi batch reactor using the a-jumping gene adaptations of genetic algorithm and simulated annealing, *Polym Eng Sci* 48:2198–2215, 2008.
Ramteke M, Gupta SK: Multi-objective genetic algorithm and simulated annealing with the jumping gene adaptations. In Rangaiah GP, editor: *Multi-objective optimization: techniques and applications in chemical engineering*, Singapore, 2009a, World Scientific, pp 91–129.
Ramteke M, Gupta SK: Biomimetic adaptation of the evolutionary algorithm, NSGA-II-aJG, using the biogenetic law of embryology for intelligent optimization, *Indus Eng Chem Res* 48:8054–8067, 2009b.
Ramteke M, Gupta SK: Biomimicking altruistic behavior of honey bees in multi-objective genetic algorithm, *Indus Eng Chem Res* 48:9671–9685, 2009c.
Ray WH, Szekely J: *Process optimization with applications in metallurgy and chemical engineering,* New York, 1973, Wiley.
Reklaitis GV, Ravindran A, Ragsdell KM: *Engineering optimization,* New York, 1983, Wiley.
Ripon KSN, Kwong S, Man KF: Real-coding jumping gene genetic algorithm (RJGGA) for multi-objective optimization, *Inf Sci* 177:632–654, 2007.
Sankararao B, Gupta SK: Multi-objective optimization of an industrial fluidized-bed catalytic cracking unit (FCCU) using two jumping gene adaptations of simulated annealing, *Comp Chem Eng* 31:1496–1515, 2007a.
Sankararao B, Gupta SK: Multi-objective optimization of pressure swing adsorbers for air separation, *Indus Eng Chem Res* 46:3751–3765, 2007b.
Sharma S, Nabavi SR, Rangaiah GP: Performance comparison of jumping gene adaptations of elitist non-dominated sorting genetic algorithm. In Rangaiah GP, Bonilla-Petriciolet A, editors: *Multi-objective optimization: developments and prospects for chemical engineering*, New York, 2013, Wiley in press.
Sikarwar GS: Array informatics: robust clustering of cDNA microarray data. M.Tech. Thesis, Indian Institute of Technology, Kanpur, 2005.
Simoes AB, Costa E: Transposition vs. crossover: an empirical study. In *Proc. of GECCO-99*, Orlando, FL, 1999a, Morgan Kaufmann, pp 612–619.
Simoes AB, Costa E: Transposition: a biologically inspired mechanism to use with genetic algorithm. In *Proc. of the 4th ICANNGA99*, Berlin, 1999b, Springer, pp 178–186.
Stryer L: *Biochemistry,* ed 4, New York, 2000, W. H. Freeman.

Wajge RM, Gupta SK: Multiobjective dynamic optimization of a nonvaporizing nylon 6 batch reactor, *Polym Eng Sci* 34:1161–1172, 1994.
Wajge RM, Rao SS, Gupta SK: Simulation of an industrial semibatch nylon 6 reactor: optimal parameter estimation, *Polymer* 35:3722–3734, 1994.
Wongsu F, Hidajat K, Ray AK: Application of multi-objective optimization in the design of simulated moving bed for chiral drug separation, *Biotechnol Bioeng* 87:704–722, 2004.
Wright AH: Genetic algorithms for real parameter optimization. In Rawlins GJE, editor: *Foundations of genetic algorithms 1 (FOGA 1)*, Orlando, FL, 1991, Morgan Kaufmann, pp 205–218.
Yu W, Hidajat K, Ray AK: Application of multi-objective optimization in the design and operation of reactive SMB and its experimental verification, *Indus Eng Chem Res* 42:6823–6831, 2003.
Yuen CC, Aatmeeyata, Gupta SK, Ray AK: Multiobjective optimization of membrane separation modules using genetic algorithm, *J Membrane Sci* 176:177–196, 2000.
Zhang Z, Hidajat K, Ray AK: Multi-objective optimization of simulated moving bed reactor for MTBE synthesis, *Indus Eng Chem Res* 41:3213–3232, 2002a.
Zhang Z, Hidajat K, Ray AK, Morbidelli M: Multiobjective optimization of SMB and varicol process for chiral separation, *AIChE J* 48:2800–2816, 2002b.
Zhang Z, Mazzotti M, Morbidelli M: Multiobjective optimization of simulated moving bed and varicol processes using a genetic algorithm, *J Chromatogr A* 989:95–108, 2003a.
Zhang Z, Mazzotti M, Morbidelli M: Powerfeed operation of simulated moving bed units: changing flow-rates during the switching interval, *J Chromatogr A* 1006:87–99, 2003b.
Zhang Z, Morbidelli M, Mazzotti M: Experimental assessment of power feed chromatography, *AIChE J* 50:625–632, 2004.

INDEX

Note: Page numbers followed by "*f*" indicate figures, and "*t*" indicate tables.

I

J

K

L

M

S

X

Z

CONTENTS OF VOLUMES IN THIS SERIAL

Volume 33 (2007)

Shinichi Matsumoto and Hirofumi Shinjoh, *Dynamic Behavior and Characterization of Automobile Catalysts*

Mehrdad Ahmadinejad, Maya R. Desai, Timothy C. Watling and Andrew P.E. York, *Simulation of Automotive Emission Control Systems*

Anke Güthenke, Daniel Chatterjee, Michel Weibel, Bernd Krutzsch, Petr Kočí, Miloš Marek, Isabella Nova and Enrico Tronconi, *Current Status of Modeling Lean Exhaust Gas Aftertreatment Catalysts*

Athanasios G. Konstandopoulos, Margaritis Kostoglou, Nickolas Vlachos and Evdoxia Kladopoulou, *Advances in the Science and Technology of Diesel Particulate Filter Simulation*

Volume 34 (2008)

C.J. van Duijn, Andro Mikelić, I.S. Pop, and Carole Rosier, *Effective Dispersion Equations for Reactive Flows with Dominant Peclet and Damkohler Numbers*

Mark Z. Lazman and Gregory S. Yablonsky, *Overall Reaction Rate Equation of Single-Route Complex Catalytic Reaction in Terms of Hypergeometric Series*

A.N. Gorban and O. Radulescu, *Dynamic and Static Limitation in Multiscale Reaction Networks, Revisited*

Liqiu Wang, Mingtian Xu, and Xiaohao Wei, *Multiscale Theorems*

Volume 35 (2009)

Rudy J. Koopmans and Anton P.J. Middelberg, *Engineering Materials from the Bottom Up – Overview*

Robert P.W. Davies, Amalia Aggeli, Neville Boden, Tom C.B. McLeish, Irena A. Nyrkova, and Alexander N. Semenov, *Mechanisms and Principles of 1 D Self-Assembly of Peptides into β-*Sheet Tapes

Paul van der Schoot, *Nucleation and Co-Operativity in Supramolecular Polymers*

Michael J. McPherson, Kier James, Stuart Kyle, Stephen Parsons, and Jessica Riley, *Recombinant Production of Self-Assembling Peptides*

Boxun Leng, Lei Huang, and Zhengzhong Shao, *Inspiration from Natural Silks and Their Proteins*

Sally L. Gras, *Surface- and Solution-Based Assembly of Amyloid Fibrils for Biomedical and Nanotechnology Applications*

Conan J. Fee, *Hybrid Systems Engineering: Polymer-Peptide Conjugates*

Volume 36 (2009)

Vincenzo Augugliaro, Sedat Yurdakal, Vittorio Loddo, Giovanni Palmisano, and Leonardo Palmisano, *Determination of Photoadsorption Capacity of Polychrystalline TiO2 Catalyst in Irradiated Slurry*

Marta I. Litter, *Treatment of Chromium, Mercury, Lead, Uranium, and Arsenic in Water by Heterogeneous Photocatalysis*

Aaron Ortiz-Gomez, Benito Serrano-Rosales, Jesus Moreira-del-Rio, and Hugo de-Lasa, Mineralization of Phenol in an Improved Photocatalytic Process Assisted with Ferric Ions: *Reaction Network and Kinetic Modeling*

R.M. Navarro, F. del Valle, J.A. Villoria de la Mano, M.C. Alvarez-Galván, and J.L.G. Fierro, *Photocatalytic Water Splitting Under Visible Light: Concept and Catalysts Development*

Ajay K. Ray, *Photocatalytic Reactor Configurations for Water Purification: Experimentation and Modeling*

Camilo A. Arancibia-Bulnes, Antonio E. Jiménez, and Claudio A. Estrada, *Development and Modeling of Solar Photocatalytic Reactors*

Orlando M. Alfano and Alberto E. Cassano, *Scaling-Up of Photoreactors: Applications to Advanced Oxidation Processes*

Yaron Paz, *Photocatalytic Treatment of Air: From Basic Aspects to Reactors*

Volume 37 (2009)

S. Roberto Gonzalez A., Yuichi Murai, and Yasushi Takeda, *Ultrasound-Based Gas–Liquid Interface Detection in Gas–Liquid Two-Phase Flows*

Z. Zhang, J. D. Stenson, and C. R. Thomas, *Micromanipulation in Mechanical Characterisation of Single Particles*

Feng-Chen Li and Koichi Hishida, *Particle Image Velocimetry Techniques and Its Applications in Multiphase Systems*
J. P. K. Seville, A. Ingram, X. Fan, and D. J. Parker, *Positron Emission Imaging in Chemical Engineering*
Fei Wang, Qussai Marashdeh, Liang-Shih Fan, and Richard A. Williams, *Electrical Capacitance, Electrical Resistance, and Positron Emission Tomography Techniques and Their Applications in Multi-Phase Flow Systems*
Alfred Leipertz and Roland Sommer, *Time-Resolved Laser-Induced Incandescence*

Volume 38 (2009)
Arata Aota and Takehiko Kitamori, *Microunit Operations and Continuous Flow Chemical Processing*
Anıl Ağıral and Han J.G.E. Gardeniers, *Microreactors with Electrical Fields*
Charlotte Wiles and Paul Watts, *High-Throughput Organic Synthesis in Microreactors*
S. Krishnadasan, A. Yashina, A.J. deMello and J.C. deMello, *Microfluidic Reactors for Nanomaterial Synthesis*

Volume 39 (2010)
B.M. Kaganovich, A.V. Keiko and V.A. Shamansky, *Equilibrium Thermodynamic Modeling of Dissipative Macroscopic Systems*
Miroslav Grmela, *Multiscale Equilibrium and Nonequilibrium Thermodynamics in Chemical Engineering*
Prasanna K. Jog, Valeriy V. Ginzburg, Rakesh Srivastava, Jeffrey D. Weinhold, Shekhar Jain, and Walter G. Chapman, *Application of Mesoscale Field-Based Models to Predict Stability of Particle Dispersions in Polymer Melts*
Semion Kuchanov, *Principles of Statistical Chemistry as Applied to Kinetic Modeling of Polymer-Obtaining Processes*

Volume 40 (2011)
Wei Wang, Wei Ge, Ning Yang and Jinghai Li, *Meso-Scale Modeling—The Key to Multi-Scale CFD Simulation*
Pil Seung Chung, Myung S. Jhon and Lorenz T. Biegler, *The Holistic Strategy in Multi-Scale Modeling*
Milo D. Meixell Jr., Boyd Gochenour and Chau-Chyun Chen, *Industrial Applications of Plant-Wide Equation-Oriented Process Modeling—2010*
Honglai Liu, Ying Hu, Xueqian Chen, Xingqing Xiao and Yongmin Huang, *Molecular Thermodynamic Models for Fluids of Chain-Like Molecules, Applications in Phase Equilibria and Micro-Phase Separation in Bulk and at Interface*

Volume 41 (2012)
Torsten Kaltschmitt and Olaf Deutschmann, *Fuel Processing for Fuel Cells*
Adam Z.Weber, Sivagaminathan Balasubramanian, and Prodip K. Das, *Proton Exchange Membrane Fuel Cells*
Keith Scott and Lei Xing, *Direct Methanol Fuel Cells*
Su Zhou and Fengxiang Chen, *PEMFC System Modeling and Control*
François Lapicque, Caroline Bonnet, Bo Tao Huang, and Yohann Chatillon, *Analysis and Evaluation of Aging Phenomena in PEMFCs*
Robert J. Kee, Huayang Zhu, Robert J. Braun, and Tyrone L. Vincent, *Modeling the Steady-State and Dynamic Characteristics of Solid-Oxide Fuel Cells*
Robert J. Braun, Tyrone L. Vincent, Huayang Zhu, and Robert J. Kee, *Analysis, Optimization, and Control of Solid-Oxide Fuel Cell Systems*

Volume 42 (2013)
T. Riitonen, V. Eta, S. Hyvärinen, L.J. Jönsson, and J.P. Mikkola, *Engineering Aspects of Bioethanol Synthesis*
R.W. Nachenius, F. Ronsse, R.H. Venderbosch, and W. Prins, *Biomass Pyrolysis*
David Kubička and Vratislav Tukač, *Hydrotreating of Triglyceride-Based Feedstocks in Refineries*

Tapio Salmi, *Chemical Reaction Engineering of Biomass Conversion*
Jari Heinonen and Tuomo Sainio, *Chromatographic Fractionation of Lignocellulosic Hydrolysates*

Volume 43 (2013)

Grégory Francois and Dominique Bonvin, *Measurement-Based Real-Time Optimization of Chemical Processes*
Adel Mhamdi and Wolfgang Marquardt, *Incremental Identification of Distributed Parameter Systems*
Arun K. Tangirala, Siddhartha Mukhopadhyay, and Akhilananand P. Tiwari, *Wavelets Applications in Modeling and Control*
Santosh K. Gupta and Sanjeev Garg, *Multiobjective Optimization Using Genetic Algorithm*

FSC
www.fsc.org
MIX
Papier | Fördert gute Waldnutzung
FSC® C083411

Kolibri 360 GmbH
Ferdinand-Jühlke-Straße 7
99095 Erfurt, Deutschland
produktsicherheit@kolibri360.de